YOUNG SCIENTIST USA.
SOCIAL SCIENCE

Copyright © 2014 by Authors

Layout by Gilbert Rafanan, Jeya Jeevan, Irene Cruz and Ruslan Nabiev
Cover © by Eugene Shishkov

All rights reserved. In accordance with the U.S. Copyright Act of 1976, the scanning, uploading, and electronic sharing of any part of this book without the permission of the publisher constitute unlawful piracy and theft of the author's intellectual property. If you would like to use material from the book (other than review purposes), prior written permission must be obtained by contacting the publisher.

The publisher is not responsible for websites (or their content) that are not owned by the publisher.

3702 W Valley HWY N
STE 204-31245
Auburn, WA
98001

http://www.YoungScientistUSA.com/

Printed in the United States of America

Lulu, 2014

ISBN 978-1-312-13261-0

Table of Contents

Economics and Business

The System of Regulation of the Oil Complex of the Russian Federation 6
Raisa Azieva

Technologies Applied in Micro-financing. 11
Nazim L. Balamirzoyev

The Analysis of Influence of the Global Financial Crisis on Introduction of
Information Technologies in the Russian Banking Sector 14
Alexey Bataev

International Competitive Ability of the National Economy of Azerbaijan 18
Asaf G. Garibov

Return and Risk — Basic Indicators of the Quality of Equity Securities for Investors. 22
Natalia Gorbunova and Oksana Kavkaykina

Problems of Evaluation of Business Value in Conditions of
Cyclic Evolution of Economic Systems. 26
Tatiana Guseva

The Social Policy and Its Features in the Conditions of Transition to Market Relations 30
Roza Karajanova and Gulzar Uralbaeva

Strategy of Development of the Primary and Secondary Vocational
Education in the Russian Federation 34
Sofia Morozova and Natalya Chernyshova

Capital Management Through the Lense of Economical Model of Timing Cycles 43
Andrey Nesterov

Theoretical Bases of the Aggregated Condition of Innovative Process. 47
Evgenii Plakhin

Monetary Theory from Neoclassicists to Monetarism (Evolution Aspect) 51
Anna Sedova and Anna Ratzlaf

Development of the Concept of the Optimum Mechanism of
Regional Government by Economy of Agro-Industrial Complex. 55
Askar Sharipov and Gabit Ahmetzhanov

Public Efficiency of the Energy Market: Concept and Measurement. 61
Mikhail A. Simonov

Axiological Approach to Application of Innovations in Work of the
Small Commercial Enterprise. 67
Maxim A. Titovets

Application of Balanced Scorecard System in the Environmental Economics 76
Marina Treyman

Cardinal Voting: the Way to Escape the Social Choice Impossibility 80
Sergei A. Vasiljev

Construction Waste Management in the Republic of Bulgaria
(Legislative and Practical Aspects) 86
Gena Tsvetkova Velkovska

Law

International and National Experience of Application of Ecological Audit 96
Victoria Erofeeva and Vasilisa Kraeva

Commercial Arbitration as a Civil Society Institution 98
Tatyana Letuta

Achieving Social Consensus as a Strategic Goal of the Russian State 103
Aleksey Treskov

Political Science

The Partnership of the Republic of Uzbekistan with UN for the
Problems of Central Asian Regional Security 111
Oybek Abdimuminov

The Standards of Coordinating the Benefits of the Centre and the
Provinces in Uzbekistan 115
Ruzimat Juraev

Kazakh Enlighteners' Views on Counseling Political and Social Problems 117
Murat O. Nassimov, Bauyrzhan Akhmetov, and Botagoz Paridinova

International Terrorism: Definition, Essence, Main Features 121
Alexander Pokhilko

Psychology

The Impact of Meditation on Emotional Intelligence of Migrants as a
Key Factor of Social-Psychological Adaptation 126
Sergey V. Afanasyev

Music as an Effective Method of Teaching English as a
Foreign Language at School 133
Maria V. Arkhipova

Genius Sleeps inside Everyone, or What Is Real Genius: Pathology or Destiny
and How to Become a Genius 137
Anna Dukhareva and Alexey Mayorov

New Approach for Integration of Psychological Knowledge. 141
Anton Karasevich

The Problems of Formation of Social-Psychological Adaptation and
Communicative Competence in Students of Higher Education Institutions. 147
Guli S. Salomova

About Relationship Between Defense Mechanisms Of The Mentality and
The Level Of School Anxiety Of Senior Pupils. 152
Natalia Turan

Sociology

Impact of the Demographic Structure of the Population on the Stability
of an Urban Family 156
Fania Igebaeva

The Specific Features of Uzbek People's Ethnoculture 159
Maksuda S. Khajieva, Mangubek Urazmetov, and Sayyora Akmanova

Understanding of Charity in Russia and its Interaction with the State. 162
Maria Kislyakova

Political Consciousness of the Youth as a Foundation for the
Development of Civil Society 167
Eleanora Yusupova

ECONOMICS AND BUSINESS

The System of Regulation of the Oil Complex of the Russian Federation

Raisa Azieva

Grozny State Oil Technical University, Grozny, Russia

Abstract. *This work describes methods of effective state regulation of the fuel and energy complex in the area of regulation and development of the oil complex and its infrastructure, and also concerns issues of formation of prices for oil and oil products sold on the domestic market. The author formulates various aspects requiring legislative solutions, which are connected with the delineation of authority at different levels.*

Keywords: *oil complex, state regulation, oil and oil products, oil export, formation of prices, taxation policy.*

In the modern day Russian economy, the fuel and energy complex holds approximately 30 % of the volume of industrial production, 32 % of the income of the consolidated budget and 54 % of the federal budget, 54 % of exports, approximately 45 % of currency earnings and also full provision of domestic needs for liquid fuel. That is why special attention is paid to effective state regulation of this segment of the Russian economy. State regulation is carried out in two main directions:

- regulation of the functioning and development of the oil complex and its infrastructure;
- regulation of issues of formation of prices for oil and oil products, effected on the domestic market.

Historically the results of development of the oil industry influence the development of the Russian economy and this is especially apparent under conditions of globalization. That is why any rapid changes on the global oil market have a severe negative impact on our economy. Equally important are domestic prices for oil and oil products. It is not a secret that there are not many goods that can be competitive on the international market, and those that can, do so due to low internal prices of energy and fuel. In this area there is a direct correlation between changes of internal prices of oil products and prices of other kinds of goods. So the dynamics of internal prices are as important to the Russian economy as the dynamics of international prices. And so under conditions of low-paying capacity of the population, the rise in prices is an important problem for the government. Because the prices in the Russian oil market in recent years have only had the tendency to increase, only the pace of this increase is changing. At the present moment prices have reached their limit, and the choice of the government and oil companies' policy in this area will determine the further fate of the domestic oil complex.

An explanation of the rise of internal prices based on the direct influence of the global oil market with its upward price dynamics would seem correct and logical if Russia didn't produce and refine oil itself, and was instead satisfying its needs by importing oil products. The reduction of crude oil exports is the main factor influencing the dependency of domestic prices on the outer business environment; for this reason global trends influence

Russia's domestic market. The growth of export in volume leads to a relative deficit of oil products on the domestic market and, consequently, to the rise of domestic prices.

This type of situation characterized the Russian market's oil products in the summer of 2000, when there was a temporary cancellation of domestic market supply; after a mandatory reduction in household supplies, the selling prices of domestic oil refineries made a rapid «reach» towards global prices and retail prices almost reached the upper limit of the purchasing capacity of Russian car owners.

Many experts specify two factors: protecting the Russian market from the influence of the global price rise for oil, and oil products themselves — their low quality and expensive logistics.

Other researchers think that the level of domestic retail prices is determined not only by the current business environment of the fuel market, but also by relatively stable relations among the varying utility of particular oil products. By competing with alternative sources and kinds of energy (gas, coal, electric energy, etc.) and other (non-energy) needs of potential consumers, retail prices will sooner or later fit into their own price niche, reflecting their relative deficiency and effectiveness of use as compared with other material and nonmaterial benefits. That is why sudden bursts of prices for Russian automobile fuel, occurring from time to time, are not directly connected with global price growth for oil and oil products, but reflect temporary trade deficits on the domestic fuel market.

We should also note the fact that due to the rise in prices for petroleum in Russia, the behavior of consumers is also changing. Consumers' demand, once at a stable rate of increase in car possession by 6–8 %, is gradually becoming elastic in price.

Therefore we can draw the following conclusions:

- the rise in prices for petroleum in Russia will be limited by a deficit of export capacity and the gradual growth of price elasticity of domestic demand;
- it is obvious that clear state policy in the field of formation of prices for oil products, designated for the domestic market, must be developed, because the methods of formation of prices used by oil companies have no real correlation with factors of the domestic and international environments, and are strictly speculative;
- the rise of prices on the domestic market cannot be explained by the rise of global prices for oil, because oil companies are not buying it there, so their expenditures for purchasing raw materials are determined only by their own expenses for oil extraction.

The current system of regulation of the oil market, although it does not always function effectively, can become the basis for building a new one, because it includes a number of mechanisms, which, used together, must provide for effective functioning of the oil complex. Elements of control of the development of infrastructure are partially included in this system, but primarily on a federal level. We can distinguish the following basic elements:

- regulation of subsurface resources management relations;
- agreements on sharing production;
- state regulation of investment processes in the Russian oil industry;
- tax policy in the oil sector.

When developing natural resources, global practice has determined such methods of subsurface resources management as a contract form of agreements with the investor, licenses as a form of administrative right and concessions as a form of civil law.

The main characteristic of any agreement system is the fact that the contract is concluded between company-supplier and state, and thus their relations move into the territory of civil law. If the state can withdraw a license unilaterally, the contract will not allow that. The state also cannot change particular articles of the contract, the most important of them being fixing of the tax burden for the entire period of its validity. With a system of licenses, the tax burden on the licensee can change, following changes in tax laws. All developed countries prefer a system of licenses in the field of subsurface resources management. This form of subsurface resources management is used in Norway, Australia, USA, Great Britain, Canada and other developed countries.

A Production Sharing Agreement (PSA) specifies that only part of the raw materials extracted becomes the property of the investor company. The other part is passed over to the receiving party as a payment for the use of natural resources. All risk of exploration lies on the investor, because associated expenditures will be covered only in the case that commercially beneficial resources are found in the exploration location. If such resources are not found, the area of land is returned to the state. Resource management on the basis of Production Sharing Agreement is performed in Organization of Petroleum Exporting Countries country members, Egypt, Azerbaijan and Mexico. The fact that the acquisition of income in the form of oil allows the minimization of currency risks makes this approach the most effective in developing countries and countries with a transitional economy.

In Russia at the present moment the system of subsurface resource management is two-fold, which makes it rather ambiguous - we have:

- the license system, based on administrative law, according to the law «Concerning Subsurface Resources», put into force in 1992;
- the contractual system, based on civil law, put into force in 1995, according to the law «Production Sharing Agreement».

27,756 licenses were issued in Russia in 2011, and half of them have already ceased to be in force. 91.2 % of explored or extracted oil resources and 83 % of gas resources were handed over to license owners, which later violated the terms of the agreement; this proves that the license system is not effective.

Current Russian law provides the right to use subsurface resources only in the following cases: either by the decision of authorized executive authorities or by a production sharing agreement. Provision of subsurface resources for use in all cases is formalized by licenses. This means that regulation of subsurface resources management is carried out with an administrative principle. However part 2 of article 9 of the Constitution of the Russian Federation allows various forms of ownership of land and other natural resources, including private property. The Civil Code of the Russian Federation in article 130 relates sites of subsurface resources and sites of land as being in the category of real estate, which can participate in civil transactions. There is no final and definite decision on ownership of natural resources at the present time.

In international practice both administrative and civil forms of subsurface resource use are implemented, when sites of subsurface resources are sold and leased. In countries with an Anglo-American legal system, the ownership of subsurface resources is derived from ownership of land. For example, in the USA the owner of a land site is also the owner of its natural resources, contained under its surface. The owner of a land site pays tax for the property, calculated on the basis of the cost of this land site with the exclusion of the cost of natural resources. This approach stimulates more complete extraction of liquid fuel from under the surface rather than extensive expansion of oil extraction territories. Extraction companies are not interested in increasing proven resources, because they'll have to pay additional taxes for them. That is why the average proven resources provision of leading international oil and gas companies does not extend beyond 10–15 years, whereas for Russian companies it is often 50 years and more. According to data from the Presidential Chief Supervisory Department, raw material supply provision for the «LUCOIL» oil company amounts to 33 years, for the «YUKOS» oil company— 47 years, for TNK — 80 years. Such provision of resources allows these companies to extract only 40–50 % of resources; after that they «leave» the oil wells.

Another question, demanding a legislative solution, is connected with the delineation of authority at different levels. The Constitution of the Russian Federation relates questions of use and administration of subsurface resources to the joint competence of the Russian Federation and its constituent entities. However this differentiation cannot be found expressed in any documents in a clear manner. The Principle of «two keys», included in the «Concerning Subsurface Resources» 1992 law (when license for ownership of subsurface resources is fixed with two signatures — by a representative of the Ministry of Nature and by the head of the Russian Federation constituent entity, in which the subsurface resources are located) creates a lot of misunderstanding and even legal proceedings, and leads to irresponsibility and corruption.

For this reason we should start by getting rid of this duality in management of subsurface resources, and spread authority between state government bodies of the Russian Federation and constituent entities of the Russian Federation.

The Federal law «Production Sharing Agreement» was signed on 30th of December 1995. According to current law, the scheme of carrying out this agreement is as follows: all manufactured products are obligatorily divided into three parts:

- compensating production, designated to the investor as compensation for expenses on conduction of work according to the agreement;
- profit production is divided between investor and state;
- part of profit production of the investor are his payments for use of subsurface resources — a royalty, paid to the state.

Since the basic law was accepted by parliament, 29 oil deposits have been set up to be explored on the basis of PSA. At the present moment three agreements are being carried out; all of them were concluded before this law came into legal force. The reasons why Russian and international companies have not yet concluded a single PSA in 8 years lie, in the first place, in this law itself.

Because PSAs are concluded on the basis of agreements and finding compromises between investor and state, the economic effect is estimated for each of these sides; this is the basis on which they decide whether to enter into this agreement or not. But as the main state strategy towards PSA at the present moment is the provision of budget, this regime becomes unattractive to the investors and they do not invest monetary assets into projects.

Therefore approaches to determining state interests in agreeing a PSA need to be revised. Because benefits to the state from a PSA cannot be estimated only in terms of tax incomes, it will also receive oil, and the budget will be filled by income from its realization. Moreover, PSA parameters can be chosen in such a way that incomes of state and investor are almost equal to values of these parameters under the current tax system.

The tax policy of the oil sector is the sphere where state interests and the main principles of regulation are carried out in reality. Under the present system of taxation in the oil sector, included are both a general tax base for all economic entities, and individual taxation systems, provided, for example, within the scope of production sharing agreements. The main taxes are: VAT and export duties (that is taxes, excluded from gross revenues when calculating net profits); extraction tax (turnover tax); tax for property of organizations; contributions to non-state non-budgetary funds; personal income tax; tax on the profit (income) of organizations.

Despite the increase in the tax burden on oil companies, incomes of the oil sector continue to increase, following the rise of prices for oil. An explanation for this lies in the active use of tax burden optimization schemes by the companies. The most active tax planning was demonstrated by «Sibneft», «YUKOS», «LUCOIL» and «TNK» companies. Net profit of «Sibneft» and «TNK» is formed outside of Russian tax jurisdiction, while the highest tax burden lies on the «Surgutneftegaz» company.

The relatively low tax burden of Russian oil-extracting companies is marked by leading international organizations. For example, the International Monetary Fund confirms the fact that the «tax burden» in the Russian oil sector is actually relatively low, as compared to other main oil extracting countries.

On the basis of the above-mentioned facts, we can draw the following conclusions:

- in Russia there is a developed multidimensional system of state regulation of the oil complex.
- special attention in the regulation of the oil complex is paid to questions of resource distribution and the withdrawal of income from oil extraction.
- regulation of development of infrastructure of the oil complex is carried out basically at state level; the regional component is almost undeveloped.
- there is no clear state policy in the area of the formation of prices for oil and oil products, which negatively impacts the country's economy as a whole.
- a legal base in the area of state planning of development of infrastructure is absent.

- tax policy in the area of oil companies' taxation is not rational, therefore the state has no sources for development of oil complex infrastructure.

If we analyze this data, we will see that the mechanisms of state regulation of the development of the oil complex turn out to be under-effective.

That is why we should take a more detailed look into issues, connected with rent as one of the factors, influencing the formation of oil complex infrastructure.

Export of oil and oil products is a major factor in the economic stabilization of Russia and it provides for the development of the oil complex and solving problems of maintaining and developing the material and technical base, increasing reliability and effectiveness of energy supply in Russia, which in turn determines the necessity of active participation of the state in the creation of conditions for its effective operation. An essential condition for the successful solution of the state's problems is the presence of detailed and well-developed policy in the considered sphere of the economy.

References

1. Azieva R. X. Development of control mechanism of oil complex on mesolevel/ Azieva R. X. — Monograph — Nalchik: Institute of Informatics and Problems of Regional Management of Kabardino-Balkarian Scientific Center of Russian Academy of Science, 2010
2. Egorov V., Korotetsky U. Approaching the upper limit// Expert — 2004 — № 9.
3. Krivoshekova E., Okuneva E. System of regulation of oil complex of Russia//Problems of economics — 2004 — № 7 — p.70
4. Orlov V. P. Mineral resources base of Russia under conditions of globalization of world economy//Foreign economic report — 2002 — № 5.

Technologies Applied in Micro-financing

Nazim L. Balamirzoyev
Dagestan State Technical University, Makhachkala, Russia

Abstract. Reasonable principles of investment and management in the course of a choice of technologies applied in micro-financing are considered and offered, and the principles of following micro-financial organizations in the course of the introduction of technologies are also defined.

Keywords: *information systems, micro-financial organizations, micro-financing, micro-financial technologies, financial organizations, information, technologies, clients.*

Micro-financing consists of retail financial services for small amounts of money for disadvantaged customers who have no access to traditional financial services. Micro-financing also includes corporate services for small and medium businesses.

Micro-financing technologies make the provision of a full range of financial services to disadvantaged small and financially sound micro businesses. Unlike micro-financing, traditional crediting does not actively work with such customers.

Micro-financing institutions (MFI) are oriented to emerging businesses that cannot obtain credit from banks. MFI significantly contribute to the increase of available commercial and investment resources, which are made into nonbank small and micro enterprises [1]. Micro-financing as a factor of «social entrepreneurship» implies the business is being undertaken for the solution of social problems.

Micro-financing gained its popularity when the law «On Micro-financing Activities» came into force last January. Due to this law small but costly credits began to grow rapidly. Today's situation resembles the banking boom of the first half of the 1990s. For two years, the number of MFIs in the registry of the Federal Service for Financial Markets exceeded two thousand. The Internet indignation at the extortionate interest rates under which MFIs lend money, coexists with exultant comments of mini bankers dealing with micro-crediting (the number of shadow micro-financiers is as numerous as the official ones).

MFI and its dynamics: within two years its growth has doubled from 27 billion to 42 billion rubles. Of course, it can't be compared with Russian banks: for the first nine months of 2013 the debt of customers increased by 1.84 trillion rubles. However, it should be noted that the maximum amount of the loan that MFI can issue is only 1 million rubles, and crediting usually occurs in smaller chunks: the average MFI issues 20–30 thousand rubles on consumer loans and no more than 400 thousand rubles on those of entrepreneurs.

There is a wide range of technologies providing efficiency, well-defined control over the process, transparency and attraction of new customers. However, most micro-financing institutions are not sure about what technologies to use for the highest possible benefit.

Technology used in micro-financing.

Information systems (IS) help micro-financing institutions (MFIs) to monitor and analyze their activities and keep records. Small MFIs can be operated by means of accounting programs or spreadsheets (Excel), but the majority of MFIs eventually require a commercial computer program for control of financial management, reporting and regulation. IS technologies also include mini-computers for recording information about customers, scoring methods and technologies used for communication of employees and affiliates, such as a corporate network or VSAT (wireless sharing data via satellite).

Larger MFIs and banks use nontraditional services technology, such as automated teller machines (ATM), point of sale (POS), networks (equipment in the stores, which allows the use of credit and debit cards for electronic payments and transfers) and mobile phones. These technologies allow customers to make payments, transfer money, withdraw cash and make deposits outside the office. Though new technologies are meant for reduction of maintenance costs for the poor in many countries, they have turned out to be less effective in comparison with traditional methods [2].

The advantages of new technologies are: more informed decisions, greater flexibility, lower operating costs, optimized reporting, growth of deposits, improvement of service technologies, more rural clients.

New service technologies usually require accurate planning and changes in the activity of MFI. Technologies will not solve any problems related with inefficiency of MFIs. Before you start the planning process, the organization must clearly define its mission, goals and especially procedures. Selecting and modernization of any technological system involves six clearly defined stages: project preparation, needs analysis, design, selection, implementation and management.

IFO should relate to technologies, as well as to other investments: ROI should be calculated and weighed against the total cost. Technology should benefit all users, including customers, employees and management.

The introduction of technologies must be supported by managers and stakeholders. Planning should be treated very accurately: the budget costs include not only the planned expenditures, but contingencies. Specialized consultants may be helpful as project managers and mediators in the relationship with suppliers.

The main principles of micro-financing.

1. Small and medium businesses, as well as disadvantaged populations, should be supplied not only by credits and loans, but any financial service.
2. Micro-financing is a key tool for the welfare of the population.
3. Only financially independent organizations can provide service to a significant number of disadvantaged people and entrepreneurs.
4. Micro-financing is aimed at establishing permanent local financial institutions.
5. Micro-crediting alone cannot solve the problem of improving the welfare of the poor.
6. High interest rates reduce the access of SMEs to financial services.
7. The state supports SME programs, rather than directly providing financial services.
8. Donor subsidies and state support exercised in the form of interest rate subsidies should complement private sector capital, and not replace it or compete with it.
9. A major constraint in the development of MFIs is the lack of technologies and staff.
10. MFIs should be financially and informatively transparent.

Today numerous non-banking financial institutions (such as, for example, various development funds and entrepreneurship support agencies, consumer credit cooperatives) try to qualify and retrain their staff into MFIs, enabling their profitability and economic viability. At this stage micro-crediting gave way to micro-financing: more and more organizations began to mobilize savings and use them to make loans. A market driven approach aimed at self-sufficiency and requiring no donations and donor subsidies turned out to be the most successful method.

Sometimes MFIs are created in the form of autonomous non-profit organizations and non-profit partnerships. Being quite suitable for micro-financing, autonomous non-profit organization and non-profit partnership organizations are rarely used for these purposes. This can only be explained by conservatism.

Noncommercial partnership is a membership based upon non-profit organization founded by citizens and (or) legal entities. The aim of the partnership is to achieve social goals, and hence micro-financing. The supreme governing body of the partnership is the general assembly. Registration of partnership, its accountability and control are carried out traditionally.

Technologically biased rapid change of the market is available only in partnership with other financial institutions, for example, with banks issuing plastic cards and credit programs for small and medium businesses in small towns, operators of electronic money and remittances for the development of innovative payment instruments. As a result, the future of micro-financing is closely connected with other financial market participants and theoretically we should strive for a system consisting of transparent, regulated and effective elements, each of which performs its function in the best interests of the consumer. What may be challenging are the alliances of MFIs with providers of electronic money, mobile operators and banks. The carrying out of all of the factors mentioned above can be achieved only through a scrutinized study of business and public consumption.

References

1. About micro-financial activity and the micro-financial organizations: The federal law of 2.07.2010 No. 151-FZ//Legal-reference system «Consultant Plus»: [Electronic resource] / Consultant Plus Company. — Placenta. updating 11.04.2012.
2. Andrew Meynkhart, «Control systems of information for microfinance: assessment structure» (Washington, District of Columbia: USAID project Advanced methods of micro-financing, 1999).
3. Charles Uoterfild and Nick Ramsing, «Control systems of information for the micro-financial organizations: textbook» GGAP technical tool No. 1 (Washington, District of Columbia: GGAP, 1998).

The Analysis of Influence of the Global Financial Crisis on Introduction of Information Technologies in the Russian Banking Sector

Alexey Bataev

Saint-Petersburg State Polytechnical University, Saint-Petersburg, Russia

Abstract. *There are three main areas in the article. They are: how the global financial crisis has influenced the development of information technologies in the Russian banking sector, it touches upon the Russian banks IT-expenses and changes in their structure, includes possible ways of information technologies development.*

Keywords: *the global financial crisis, information technologies, IT- budget, Russian banks, the automated banking systems.*

The financial crisis of 2008 has shaken the entire world economy, its consequences affect to the present time. The impact of the crisis was reflected in delay of growth rates of economy, led to an increase of a public debt in some countries and rising unemployment rate.

The crisis has not passed by Russia. Slowdown in industrial growth, outflow of capital, reducing government spending is the consequences of the global financial crisis, which are aggravated with internal specifics of the country.

Particularly, the financial crisis was reflected in Russian banks, so the government allocated hundreds billions rubles for support the Russian banking system. This decision allowed to smooth the financial situation and to provide a stable state of the Russian financial system.

Costs on information technologies are one of the indicators of sustainable credit institutions development. They are cut in the first place at the crisis. Data on expenditures for the informatization of the Russian banking sector are given in the Table 1.

Despite the emerging financial crisis in 2008, Russian banks have spent on information technologies record $ 33.8 billion rubles. During the acute phase of the crisis in 2009 there was a significant reduction of IT-budgets, which was approximately 20%. About 55% of all banks have cut their IT-budgets (see Chart 1).

The increase in budgets fell to banks with the state participation and large private banks, which received substantial government support.

The growth of expenses for information technologies has amounted to about 36 billion rubles in

TABLE 1. Costs of banks on introduction of information technologies

Years	Costs, billion rubles.
2006	24,3
2007	30,2
2008	33,8
2009	27,0
2010	36,0
2011	40,0
2012	42,0

2010 that has allowed asserting about reaching the pre-crisis level.

In 2011 and 2012 banks have allocated huge funds approximately 40 and 42 billion rubles, respectively, which significantly exceeded the pre-crisis indicators.

It would seem that these figures indicate overcoming the financial crisis in the Russian banking system, indeed not all of it is so clear, if we take the structure of the expenses of several banks into consideration (see Chart 2). The share of the two Russian banks: Sberbank and VTB, respectively, account for 41% and 4% of all expenditures. If you take all the banks with state participation of the top ten, their share will be about 48%. In spite of this we can say that the crisis in the Russian banking sector overcomes as a whole. Hence, b anks actively invest in information technologies. It is not only banks with state participation, but also the banks with private capital. For example, the increase amounted to TCS Bank (117,3%), Rusfinance Bank (103%), «MTS-Bank» (87%), «Kiwi Bank (76%), «Svyaznoy Bank»

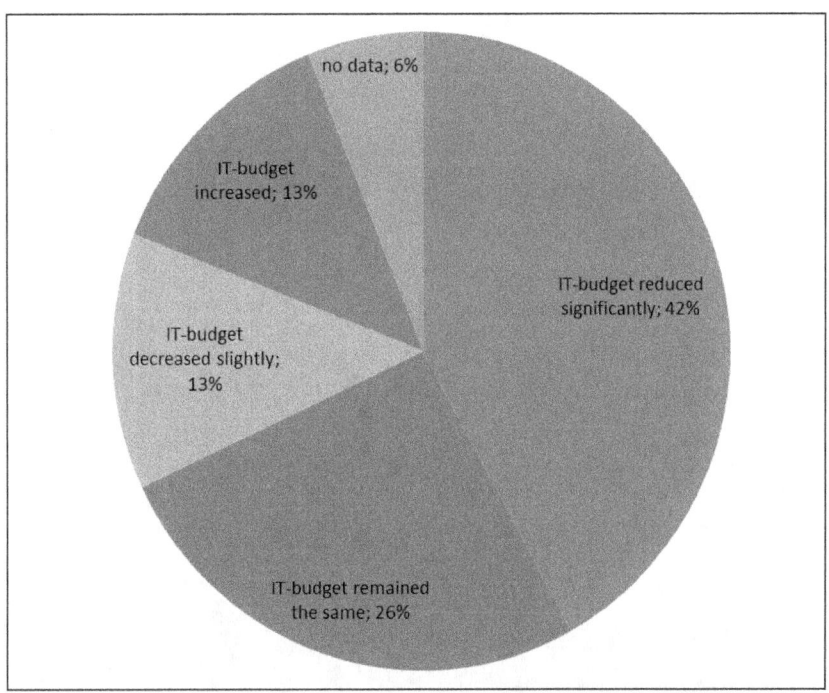

CHART 1. Reduction IT-budgets of Russian banks in 2009

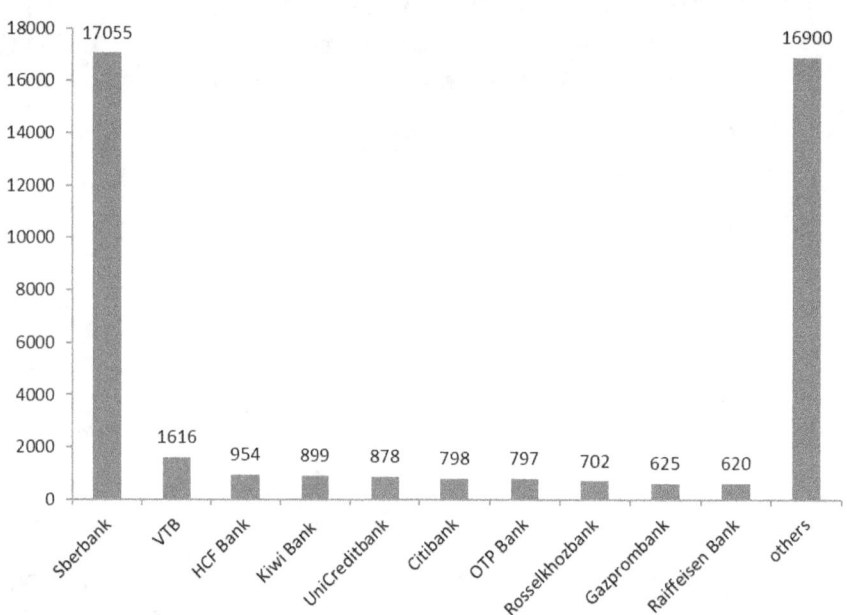

CHART 2. IT-expenses of Russian banks in 2012, million rubles

(72%), «Globex» (64%), «Binbank» (63%), «Sovcombank» (57%) in 2012 [1].

The growth of expenses for information technologies in the Russian banking sector is several times ahead of a worldwide trend. It is explained first of all that the Russian banks invest significant funds in development of the banking infrastructure, while Western banks allocate funds for its maintenance.

The first place in the structure of expenditures of Russian banks is the introduction of new software; however, the considerable share of expenses is the share of the equipment (see Chart 3) [2].

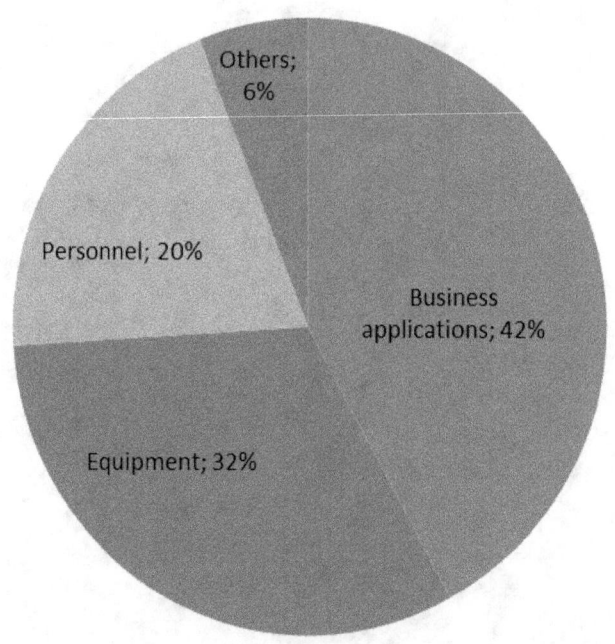

CHART 3. The structure of the Russian banks IT-expenses in 2012

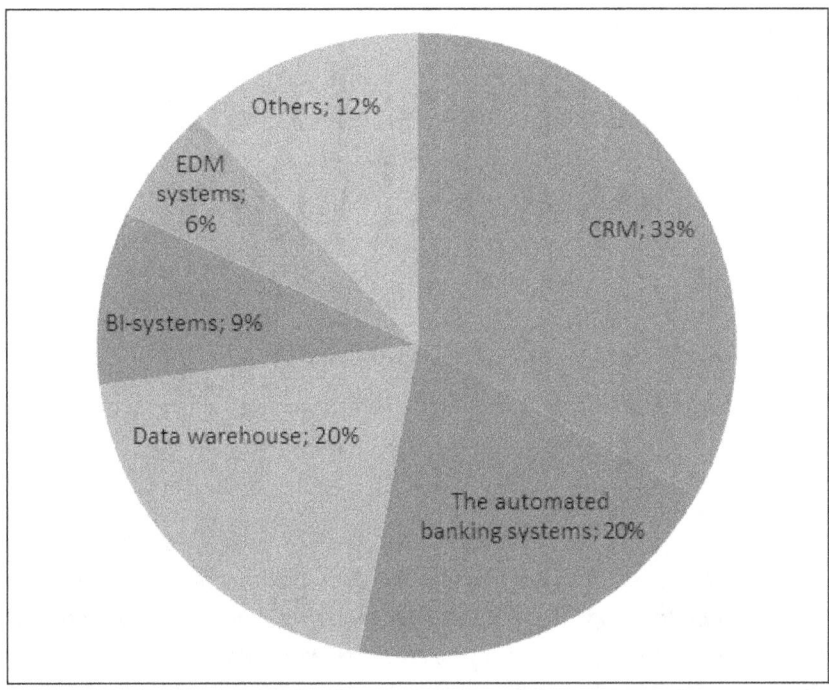

CHART 4. The priority of the IT-projects of Russian banks

The priority of projects realized in Russian banks determined on a base of the expenses structure (see Chart 4).

The introduction of CRM systems, directed on optimization of work with clients is the first priority. The introduction of these systems requires the creation of data warehouses that takes the second place.

Along with the development of systems for work with clients, a lot of banks engaged in modernization and also transition to new automated banking systems platforms. This all is evidence of the growing opportunities in the Russian banking system.

The attention is paid to the foreign automated banking systems again. In addition, it is considered the possibility of their introduction by modification with the help of Russian partners under requirements of the Russian market.

The global financial crisis has had a serious impact on the introduction of information technologies in the Russian banking sector, slowed the dynamics of development, and has led to a considerable reduction in IT-budgets during the acute phase of the crisis. Nevertheless the development of IT-technology continues. The banks have managed to restore the lost positions and are trying to catch up.

References:

1. Internet edition CNews «Russian banks exceeded 10 times world figures on the growth of IT-expenses," 2012.
2. Structure of Russian banks IT-costs in 2012, «Computerworld Russia», № 13, 2013
3. Bataev A.V. Impact of the financial crisis on the introduction of information technologies in the Russian banking sector, Proceedings of the 11th International Scientific and Practical Conference "Financial Problems of the Russian Federation and their solutions: Theory and Practice" St. Petersburg, 2010, p. 173-178
4. Bataev A.V. Trends and prospects of development of the information technologies market in the banking sector in Russia // Young scientist 2013, № 10, p. 268-271.

International Competitive Ability of the National Economy of Azerbaijan

Asaf G. Garibov
Nakhchivan State University, Nakhchivan, Azerbaijan

Abstract. *This article provides clarification of the theoretical and methodological background of the economic security and competitive ability of a national economy in view of modernization, based on system analysis. The author has specified a complex approach to estimation of how innovative responsiveness of a national economy is created, and factors limiting competitive ability; system of metrics for analysis of innovative activity is constructed.*

Keywords: *competitive ability, innovative process, modernization, innovative economy, competitive ability rating.*

Intensive development of international economic relations is one of the key factors in successful development and functioning of cooperating countries. Reaching key goals of the strategy for the social and economic development of Azerbaijan by 2020 is possible only if we use new opportunities for scientific and technical progress and qualitative technological modernization. To understand this thesis, we should thoroughly study the experience of other countries in the development of science, technique and technology and make a number of international comparisons. Questions of estimation of innovative potential abroad are periodically studied by the Council on Competitiveness (U.S.), the Institute for Strategy and Competitiveness at Harvard University (U.S.), the International Institute for Management Development (Switzerland) and the World Economic Forum (WEF) to draw up indexes of global competitive ability [1, p.14].

Today the main goal is to reach a level such that production of non-oil industries of Azerbaijan will really be able to compete with our main exporters (Turkey, Iran, Russia, etc.). That is why we need an industrial policy that addresses all stages of manufacturing, from raw materials to final goods, in order to further promote manufactured goods on the global market or, at least, in the Commonwealth of Independent States (CIS). In other words, we need to develop those industries which focus on both domestic and foreign customers. In order to do that, the republic needs an effective investment process.

"Over the last 9 months, $16 million were invested in our economy. That is a very large amount. It is much larger than last year. If we consider the fact that this amount of investment is usually made within a 1-year period, we will realize how big investments in the Azerbaijan economy are. The majority of these investments were domestic—approximately

70 percent. The remaining part is investments made by foreign entities. I consider this balance optimal. Our economy is attractive to foreign investors, but domestic investments are still, above all, a sign of an extremely successful governmental investment policy" —President Ilham Aliev.

As was said earlier, an effective investment process in Azerbaijan is the key to stimulating further growth of the economy, which is the main factor in creating economic security for the republic [2, p.315].

Economic security is an aggregate of factors and conditions, contributing to the independence of national economy, its stability and steadiness. It determines the ability of the economy to constantly maintain continuous realization of state economic interests, stable efficiency of economic entities, and normal conditions of the population's vital activities.

Economic security is the main qualitative characteristic of an economic system. Among the external factors of economic security of national economy of Azerbaijan we can distinguish the following:

- Beneficial climate for investments
- Effective realization of investment processes
- Decrease of growth of import dependence
- Optimal management of domestic and foreign indebtedness
- Decrease of capital drain

Analysis of the economic situation of most CIS countries has shown that factors above are not the most important as compared to domestic factors of economic security [3, p.146].

On September 3, 2013, one of the major economic organizations—the World Economic Forum—published new ratings of the competitive abilities of countries. Azerbaijan ranks 1st in the CIS region and 46th globally. As compared to the previous year, we went up 9 places. That means that the economic development and successes of Azerbaijan are connected with economic reforms. The report also published ratings of the macroeconomic situation. On that measure, Azerbaijan ranks 18th globally. Major credit rating agencies, which are currently decreasing the ratings of European countries, are increasing the credit rating of Azerbaijan.

Three main rating agencies increased our credit rating, which became possible due to the work done and reforms realized. That means that economic processes are reflected in statistical data.

The International Institute of Management Development issues an annual rating of global competitive ability for 59 countries; the ratings are based on analysis four basic parameters of economic conditions: economic performance; government efficiency; business efficiency; and the condition of the infrastructure, including technology, science and education. These factors are divided into 20 sub-factors comprising more than 300 criteria.

In the report for 2011, the ratings were headed by Hong Kong (1st place) and the U.S. (2nd place). The other 3 Asian "tigers" occupied 3rd place (Singapore), 6th place (Taiwan) and 21st place (South Korea). Sweden and Switzerland occupied 4th and 5th place. Azerbaijan went from 55th to 46th [4, p.19).

World Economic Forum (WEF) publishes the Global Competitiveness Report, which has been released annually since 2005. It is based on statistical data gathered in annual polls and by country. The Report pays serious attention to the influence of technology and innovations on global competitive ability. It includes 113 indicators, divided into 3 large groups and 8 subgroups. A consolidated index of global competitive ability is calculated, using the weighted arithmetic mean of the three indexes for each major group of indicators. Still The weight is calculated on the basis of estimation of the parameters of a multiple regression equation, in which regressand is gross domestic product per capita. Using this method, WEF specialists managed to determine the three main stages and two transitional stages of economic development for countries studied.

The first stage is called "economy, aimed at development of factors of production" and characterizes countries competing on the basis of factors of production, primarily labor and natural resources. In this case companies realize price-related competition. They sell particular products and goods, manufacturing them at low level of labor productivity. Thirty-seven countries are at this stage: African countries, two Latin American countries (Bolivia, Haiti), three CIs countries (Kirghizia, Moldova,

Tajikistan), India, Pakistan, Vietnam, Yemen, Nepal and some other countries of the "Third World".

In the second stage, the economy aims for increases in productivity and is characterized by continuous increase of production effectiveness, quality of production and decrease of expenses. In this case growth in salaries does not lead to increases in prices for products. The economies of 28 countries reached this stage (including China, Albania, Columbia, Indonesia, and Thailand, among others).

The 3rd stage, innovative economy, assumes a high standard of life and ability to maintain salaries just by creating new and unique products. The main competitive advantage in this case is new technologies and innovations. Thirty-five countries with developed economies are currently at this stage (including Japan, the U.S., the EU countries, and the 4 Asian "tigers").

When drawing the index of global competitive ability we must separate parameters, characterizing the level of technological development of economy and innovation parameters. The technological level is described by the WEF as the ability to adapt already existing technologies and the level of their availability. A special role is assigned to information and communication technologies (ICT). Authors of the index think that "ICT field has evolved into the main goal of development of technology" [5, p.7]. This point of view is not unfounded, because it is hard to imagine today's world without ICT, which allows the creation of a unified information space. Still, we shouldn't see the role of this industry as absolute, because maintaining an optimal living standard is only possible if we develop industries of material and non-material production evenly. Authors of the report choose direct foreign investments into technologies, level of technology availability and ICT parameters as indications influencing technological level. Of course, use of these parameters to characterize economies of developed countries is quite acceptable, but for countries with transitional economies we may need some additional parameters.

WEF specialists consider technological innovations the determining factor of competitive ability. In their words, "less developed countries can anyway increase labour productivity by adapting already existing technology or performing additional modernizations in industries, which have already reached innovation stage of development, but that doesn't already meet present-day conditions to maintain competitive ability. Considerable investments into research and developments, presence of research institutes, cooperation between universities and industry and effective protection of intellectual property are required" [4, p.8]. The authors of this report chose quality of education, research and development expenses, number of scientists and engineers, and information about intellectual property as key parameters characterizing innovative potential.

In general WEF experts took a detailed approach to studying the question of global competitive ability, and in the 2013-2014 report they divided the level of technological development of economy and innovative potential for the first time. The WEF also suggests the introduction of the so-called "index of stable competitive ability," in which the method used to compose the first report is slightly expanded, but these differences are not significant in relation to issues of technological development.

The EU summit of March 2000 took as its focus creation of competitive innovative economy. At the beginning of 2002 in Barcelona the European Council formulated a number of specified tasks in the field of stimulation of innovative development. In 2001 for the first time within the scope of the public opinion poll Eurobarometer, performed by the European Commission, an additional poll, called Innobarometer, was organized. The results were issued in the form of a corresponding report. In the future this poll will be administered annually.

In 2010 the Innobarometer was devoted to innovations in the state sector of economy and was carried out within 4000 European state organizations. The results showed that organizational and process innovations are spreading further among governmental structures, which results in simplification of client access to information, higher quality of satisfaction of requests and improvement of working conditions for state personnel.

(?????) saw the beginning of the EU Innovation Union initiative, aimed at improvement of innovative activity of the EU economy by 2014 and providing for creation of a unified European research space. As part of the initiative, the innovation report got a new name — "Innovation Union Scoreboard". Research is performed in order to determine threats and opportunities in the innovative sphere for particular regions and countries. European Commission specialists understand innovations in the economic sense as successful realization of ideas on the market or a modernized product, process or service. Following this interpretation, comparison of the innovative potential of countries is seen as comparison of particular criteria: innovative base, results of innovative activity (quantitative characteristics), and innovative development (qualitative characteristics) [6, p.12].

In 2009 the index of innovative development was calculated using 29 parameters, but in 2010 that number was reduced to 25 indicators, which better describe the development of national innovative systems. Nineteen parameters were taken from the previous report, 2 parameters were aggregated, and 5 new parameters were developed.

Indexes of innovative development definitely illustrate only some aspects of a complicated process. Qualitative characteristics of scientific and technical activity can only be multidimensional. No universal "integral" parameter, able to describe the situation in the science of innovations, has yet been developed yet [7, p.14].

As specialists of the European Commission report, they purposely chose simplified methods to make this index easier to understand. In our opinion there is also a good database behind the index, which allows them to use this methodology and ignore some deficiencies in data. The problem of developing regional ratings characterizing scientific and technological development is also important in international practice. On this topic, a document accompanying the innovative ratings of 2001 said: "... innovative activity has a strong regional component, so the Commission invites European regions to actively participate in benchmarking of innovative policy. In further issues of this rating of innovative activity regional dimension can be deepened due to input from these regions and increase in availability of information."

References

1. Competitiindeks: Where America stands.–Council on competitiveness Washington LC. 2007.
2. Samedzade Z. A. Stages of a long way. Azerbaijan economy in half a century, its reality and prospects. Baku, 2004, p.863
3. Tsygankov P. A. International relations. Moscow. 1996, p.215.(in Russian)
4. IMD World Competitiveness Yearbook 2011 — IMD Switzerland. Lausanne. 2011.
5. Schwab K. Salai — Martin X. The Global Competitiveness report 2011- 2012. Geneva: World Economic Forum. 2011.
6. Innovation Union Scoreboard, 2010. The Innovation Unions performance scoreboard for Research and Innovation — European Commission.2011.
7. Perani J. Benchmarking of innovative activity of European countries//Foresight № 1 (5), 2008, p.4–15.

Return and Risk — Basic Indicators of the Quality of Equity Securities for Investors

Natalia Gorbunova
Oksana Kavkaykina
Mordovian State University, Saransk, Russia

Abstract. *In article the main components of indicators of profitability and risk of share securities are considered and need of determination of profitability and risk of share securities locates at an assessment of investment appeal*

Keywords: *investor, capital, share security, risk, profitability, return, dividend.*

1. First Section

In any country one of the major directions of development of economy is implementation of the investment process, associated with attachment or involvement of investments, that is, material and financial resources in order to obtain certain benefits or income.

Considering the process of investing it is necessary to take into account that objects of investment activity can be presented by different forms: physical form (movable and immovable property); monetary and natural forms (fixed and working capital, scientific and technical products, property rights); only monetary form (deposits, shares, securities and other financial assets).

Nowadays, in addition to direct investments in real assets, indirect investments associated with the use of portfolio securities as an investment product are becoming more popular. Such situation is connected with the following features:

- high level of universality because security is mass, standard, interchangeable and it is characterized by uncertainty of price changes;
- liquidity, different from real assets and funds;
- certain proportion between return and risk, which quantitative and qualitative measurement is main problem arising in the investigation of securities market financial instruments.

Present time, from all variety of marketable securities equity securities or shares are the most popular.

Such situation is caused by that an action as the financial instrument allows coordinating interests of participants of investment process most effectively (Table 1).

The concepts of return and risk inherent in this type of securities should be considered.

In terms of absolute return, this feature of securities is that there is nominal value and market value. In turn, in accordance with D. W. Williams — M. D. Gordon's model [1], such basic economic concepts as capitalized return associated with the increase of market value and current or dividend

TABLE 1. Comparative characteristic of investment opportunities of securities

Types of securities	Interests of the issuer	Interests of the investor
Share	Possibility of long using funds from investors. Numerous increase in own capital.	Existence of a property right on part of the capital, the right to dividends. Participation in management.
Bond	Possibility of receiving means for financing of investment projects	Obtaining income, coupon or percent in due time. Independence of a financial position of the issuer
Bill of exchange	Use as the payment document for payment extension	Obtaining the income in a type of percent at the applet of a debt or sale to its bank. Use as an instrument of payment
Deposit and savings certificate	Attraction of temporarily free money for the short-term period	Obtaining income in a type of percent
Futures and other urgent contracts	Possibility of receiving arrived at contract purchase and sale in the short-term period	Possibility of receiving excess profit or existence of high risk

return, which is determined on the basis of accrued income or dividend are used to determine the relative return (Fig. 1) [2].

2. Second Section

Increase in market value occurs in the process of realization equity securities rights. Naturally at excess of sale price over purchase price, investor earns income, but at decrease of prices on the stock market he losses capital. The occurrence of capitalized return or increase in the market value has certain reasons. On the one hand, it is possible speculative boom, which, as a rule, has not objective long-term grounds. On the other hand, there is real growth of the enterprise assets as the result of investing profit in the development and expansion of manufacture that leads to corresponding increase in securities value of joint-stock company.

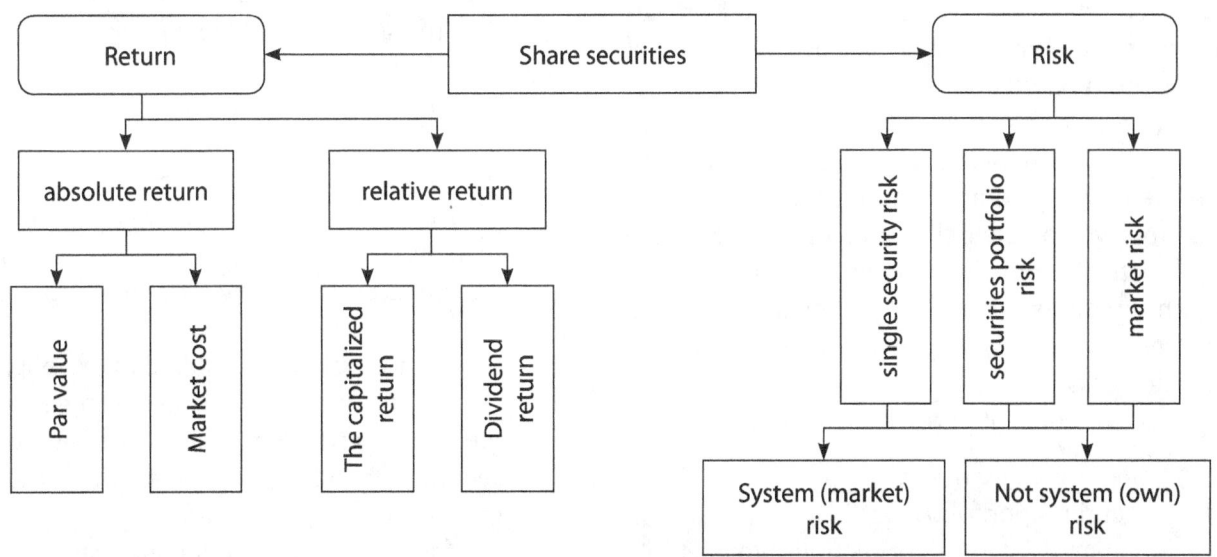

FIG. 1. Components of return and risk of share securities

Dividend is the second component of the shares return, that is, shareholder's monetary return from invested funds in the company. In this situation current return is calculated as ratio of received dividend (V) to share price existing in the market at this moment in time (C_1):

$$D_x = \frac{V}{C_1} * 100\% * \frac{T}{t} \qquad (1)$$

T — annual period, 360 days;
t — time the dividends are received.

In this direction the most important target of effective organization of joint stock company activity is to provide the optimal combination of shareholders' interests and necessity of sufficient financing of company development. The bigger part of net profit directed at payment of dividends, the smaller part remains to self-financing that leads to reduction of growth rates of owned capital, revenue and paying capacity. However, if the shareholders do not receive enough profit from invested capital and get rid of company securities, its market value will reduce and current owners will lose control over share capital.

The next qualitative characteristic of share along with return is risk. Note that the definition of return does not cause any disputes and well understood intuitively, but the definition of risk is more complicated. Taking into account the Russian economists' views, such as B.Alohina, V. Galanova, E. Mizikovsky, J.Myrkin, S. Pivovarov, A. Sheremet, risk can be described as possibility of full or partial non-fulfillment of targets set by the owner of security from based on his rights. To understand the essence of risk should know that the risk peculiar to particular securities, has a place in relation to both their aggregate and securities market in general. However, as the rights of any particular security objectively come into conflict with the rights of the other security, market risk in general is not the arithmetic sum of all components of its securities or portfolios risks.

Thus, the concept of risk exists in three forms:

a) single security risk;
b) securities portfolio risk;
c) market risk typical for securities market in general [2].

Thus, on the one hand, there is so-called system or non-diversified risk (market risk), which cannot be excluded, and it affects all securities almost equally. On the other hand there is non — system or diversified risk (unique risk), which is specific for each security and can be adjusted with effective management of securities portfolio. System risk affects all securities and it is based on factors related to economic activity and whole-market price and income volatility. The level of system risk is primary to any investor in securities market and is often connected with the inflation and interest rate risk, risk of legislation, changes in stock market and military conflicts.

Components of a non-system (equity) risk are more complex. This risk is determined by differences in investment areas in countries, regions and sectors, specificity of company activity, characteristic of each security including conditions of issue and circulation; and also qualification of operators working with securities and applied system of calculation. The main types of non — system risk are:

- risk of market prices changes. Volatility of market prices when prices at the time of transaction implementation differ from prices at the time of final calculation;
- business — risk. Low level of services, lack of modern technical solutions, high cost of infrastructure and lack of the necessary counterparty level. All these problems can lead to loss of customers and reduce business;
- operating risk. It occurs due to staff errors, organizational shortcomings, delays, fraud, systems failures, inefficient plans in case of accidents, natural disasters or terrorism, and also non-fulfillment of obligations by a third side.
- risk of investors' rights infringement. Lack of obvious and accurate information about markets makes it difficult to protect customers' rights and to receive revenue through corporate actions;
- calculation risk. Non-fulfillment of obligations by the partners for one or more payment due to problems with registration of securities, geographical location of customers, establishing minimum number of securities for transaction, conducting massive sales in a short period of time [4].

To sum up, considering economic substance of risk, it should be noted that its two parts are closely interrelated and present in any type of securities. However, for maintenance the return of equity security rational organization of dividend policy of the company is very important, but talking about risk opportunity to reduce and minimize losses plays an important role. Prevention and resolution of arising risks maintains by using various tools and methods, main of which is avoidance, keeping, transfer and risk reduction. The implementation of these methods is carried out, as a rule, by receiving additional information about choosing and results, diversification, insurance, limits and other measures.

Therefore, we can conclude that when investing in equity securities real or potential investor should assess shares in two main directions: on the one hand, to determine the interrelation between risk and return; on the other hand, to study quantitative and qualitative side of this interrelation.

Exploring return and risk of security it is necessary to consider that quality indicators are not constant and change continually depending on condition of the emitter and market in general. That is why study of return and risk of investments in equity securities based on use of certain methods of their estimation and forecasting.

References

1. V. Kovalev. Financial analysis: methods and procedures. M: Finance and statistics, 2010. 560p.
2. N. Gorbunova. Development of methods of an assessment of return and risk of share securities of the industrial enterprises. The thesis abstract on competition of a scientific degree of candidate of economic sciences. Saransk: publishing house of the Mordovian university 2005.18 p.
3. E.Mizikovsky, V. Edronova. Accounting and analysis of financial assets: shares, bonds, promissory notes. M: Finance and statistics, 2009. 289p.
4. William F. Sharp, Gordon Dzh. Alexander, Jeffrey V. Bailey. Investments. M: Infra-M, 2007. 1035p.

Problems of Evaluation of Business Value in Conditions of Cyclic Evolution of Economic Systems

Tatiana Guseva

South Ural State University, Zlatoust, Russia

Abstract: *In this article the problem of evaluation of the enterprise value in the conditions of instability of the economic system is considered. The methods of adopting the standard approaches of evaluation of enterprise value are suggested, and also the possibility to use mathematic tools of the fuzzy-set theory for evaluation is studied.*

Keywords: *evaluation of enterprise value, fuzzy-set theory.*

The financial crisis showed that the actual methods of evaluation of company's values and risks associated with the investments into business do not take into consideration the cyclic character of development of the economic system. This means that during the periods of economic boom these methods enhance the increase of value, while during economy's decline they lead to decrease of value and, consequently, to the downward shift of results of evaluation. Because of crisis the situation is changed dramatically as far as the price expectations are concerned. The data and methods accumulated during the period of fast price growth may not provide the reliable evaluation of value, so the significant changes in the methodology will be required.

In economics and in practice three basic approaches to evaluation of the enterprise value are most widely used: a) income method; b) comparative (market) method; c) cost method.

As a variant of income method which is adapted to the conditions of the fluctuating market system can be the real options method. This method allows maximum binding of evaluation to the change of situation in terms of crisis. The option has the time premium which exceeds the cash flows that can be created during the next period if the project is confirmed. Such options are called "options to delay". The option can create the difference during the enterprise evaluation in the following cases:

- the investor or the company is in possession of undervalued land;
- the enterprise holds a patent or several patents;
- the company deals with natural resources and has undervalued assets which it can use at any time as it would choose.

There are two more options which are often incorporated into the investment evaluations and effect the enterprise value:

- the option to expand investments not only due to new markets, but also due to new products in order to reach the benefit from favorable terms and conditions. This option can significantly raise the value of the

young companies to the level which is much higher than the present value of expected cash flows;
- the option to abandon investments or the option to scale down investments, which allows decreasing the risk and pulling down the bottom limit of big investments, that will result in their value becoming higher.

Within the scope of the comparative method the evaluation of the enterprise value is obtained from the known value of the comparable business, similar to the business under evaluation by a set of basic parameters. In common practice under evaluation of business value the evaluation of shares cost of the company is understood, as a rule. However, in terms of crisis it is recommended to consider other indicators. First of all, those indicators should be concerned which can be calculated on the basis of enterprise balance (financial leverage, net assets etc.). For example, the information about index of ratio between comparable companies' values and the index of efficiency of their operations (for example, earnings before interest, taxes and depreciation EBITDA) can be used. The business value under evaluation is calculated by correcting the similar index of the company under evaluation to the required rate of return. At that, the evaluation should be made based on several multipliers. The major assumption in this method is the presupposition that during crisis in the forecasted period the company's free cash flows (cash inflow minus cash outflow) will be close to zero.

The cost method is based on the fact that business value is equal to the cost of replacement of enterprise assets and liabilities. Therefore, during evaluation of the business by means of this method the accuracy of results can be low because it is often difficult to calculate the correct cost of replacement of several assets, for example, of intangible assets or specialized equipment which is not actively marketable. Besides, it would be necessary to consider the depreciation of most assets. The replacement cannot often be fulfilled physically within short terms (for example, reproduction of trademark, business reputation, relationships with clients), otherwise the only alternative is to start from the scratch and reproduce a similar business on the basis of new assets that are not depreciated. While using this method during crisis the following issues should be considered:
- while evaluating capital assets, the real asset liquidity, but not only its balance value, must be considered;
- while evaluating reserves, it is necessary to make the term-of-storage analysis of reserves such as raw materials as well as the ready-made production and make sure that the evaluation of the most significant reserves is accurate;
- as regards the analysis of accounts receivable, here the biggest debtors must be disclosed and the receivables must be classified by the periods of their creation and by the probability of their return, and further correction must be made;
- for evaluation of financial investments the state of emitters must be studied in case these investments are represented by shares of other enterprises. If the enterprise has accommodated with a loan, this article must be evaluated based on the level of the rate of return and leverage of the borrower;
- while evaluating monetary assets it is required to make sure that they are not kept on the accounts of questionable banks and are still available;
- for evaluation of credits and loans it is required to analyze terms and conditions of the loan, to assess the possibility of the creditor increasing the interest rate or accelerating the debt, or augmenting the collateral value, and, if necessary, to make correction and increase the article.
- for evaluation of accounts payable it is necessary to make sure that there are no overdue payments to key suppliers, personnel, tax authorities, and, if necessary, to build up the reserves used in case of possible sanctions [1].

Because the enterprise is developing in the course of time, the evaluations of business value are approximate and this determines the uncertainty of final evaluations. Therefore the representation of the corresponding figures of enterprise value in the form of fuzzy numbers and usage of a mathematic tool called "soft computing" seems quite natural in this case. In the simplest case the calculation

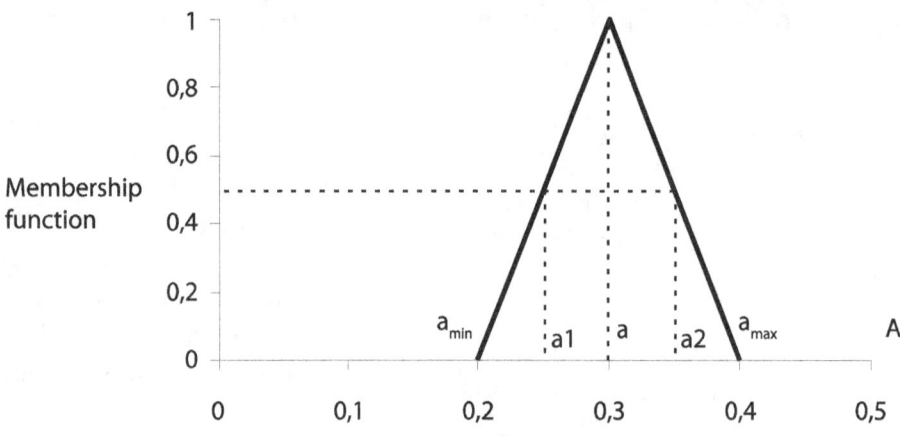

FIG. 1. Membership function of the triangular fuzzy number $\underline{A}$

of actual uncertainty can be sufficient if made just through the form of fuzzy numbers: changing the range of base set, choosing the relevant type of the membership function.

If all indicators of enterprise value are «blurred», that is, their accurate values are not known, then as the given data it would be appropriate to use the so called «triangular fuzzy numbers» with the membership function of the following type (please, see Fig. 1). By means of these numbers the assertion of the following type is modelled: «the A parameter is approximately equal to $\bar{a}$ and is unequivocally within the range $[a_{min}, a_{max}]$». In general case the fuzzy number means a fuzzy subset of the universal set of real numbers which has a normal and convex membership function [2]. Such description allows using the $[a_{min}, a_{max}]$ parameter interval and the most expected $\bar{a}$ value as the given data, then the relevant triangular figure will be $\underline{A} = (a_{min}, \bar{a}, a_{max})$.

We suggest you to use below-mentioned sets of fuzzy numbers to determine the enterprise value (Table 1).

TABLE 1. Sets of fuzzy numbers in the methods of determination of enterprise value

Approach	Method	Sets of triangular fuzzy numbers
Income	NPV-method	$\underline{r} = (r_{min}, \bar{r}, r_{max})$ – it is impossible to make an accurate evaluation of the cost of capital used in the project (for example, debt-to-equity ratio or the interest rate for long-term credits); $\underline{\Delta B_t} = (\Delta B_{min}, \overline{\Delta B_t}, \Delta B_{max})$ – the range of variation of the cash results from enterprise activities is forecasted taking into account the probable price fluctuations of salable products, the cost of consumed resources, taxation conditions and impact of other factors; $\underline{C} = (C_{min}, \bar{C}, C_{max})$ – it is impossible to represent clearly the potential conditions of the future sale of the operating enterprise or its liquidation
Comparative	Sales comparison method	$\underline{M_B} = (M_{Bmin}, \overline{M_{Bt}}, M_{Bmax})$ – it is impossible to make an accurate evaluation of the multiplier of the value of an enterprise similar by its characteristics (industry, production, technologies etc.) to the comparable objects with the known market value; $\underline{B_t} = (B_{min}, \bar{B_t}, B_{max})$ – the range of variation of the cash results from the enterprise activities is forecasted taking into account the probable price fluctuations of salable products, the cost of consumed resources, taxation conditions and impact of other factors
Cost	Net assets method	$\underline{A} = (A_{min}, \bar{A}, A_{max})$ – it is impossible to make an accurate evaluation of the market value of assets; $\underline{O} = (O_{min}, \bar{O}, O_{max})$ – it is impossible to make an accurate market value of liabilities

With the help of every fuzzy number in the structure of the given data the accuracy intervals can be specified. In case if any $\underline{A}$ parameter is known with sufficient accuracy or if it is unambiguously set, the fuzzy number $\underline{A}$ degenerates itself into the real number A with the condition $a_{min} = \overline{a} = a_{max}$ observed. At that, the essence of the method remains unchanged.

Within the framework of each of three approaches an enterprise value is a fuzzy set with the relevant membership function. The expected enterprise value is determined by transferring the argument from the relative scale on the absolute scale by the barycenter coordinate.

Thus the final evaluation of the enterprise value is determined and it will be the most expected value with due consideration of the factors of external and internal business environment of the production system.

The smaller the variation between the maximum and minimum figures of the business value is, the more sharp will become the triangle of the membership function of the enterprise value and the higher will become the confidence in calculations.

References

1. Kulakov A.B., Kulakova U.N. Some peculiarities of the business evaluation in terms of financial crisis. The Ural Social-Economic Institute, the affiliate of the Academy of Labour and Social Relations / Materials from VII Interuniversity research and practice conference, dedicated to the 100th anniversary of the sales education in the Russian Federation (2nd of March 2007) p. 176-178.
2. Sevastyanov P.V., Sevastyanov D.P. Assessment of financial parameters and investment risks from the perspective of fuzzy sets theory. Reliable programs, 1997. – № 1. – p. 10-19.

The Social Policy and Its Features in the Conditions of Transition to Market Relations

Roza Karajanova
Gulzar Uralbaeva
Karakalpakstan State University, Nukus, Uzbekistan

Abstract. *In this article a definition of social policy is suggested, and its essence and the role it carries out in the state and society is shown. The results show the special characteristics of the role of social policy at different stages in the development of the Republic of Uzbekistan. Indicators of the productivity of carrying out social policy are published in an annual report on human development. This article shows the real expenses to the republic of statistical data in the social sphere.*

Keywords: *social policy, social support, Lorentz's curve.*

One of the principles of reform of the economy of our state is the principle of carrying out strong social policy. In 1991 I.A. Karimov, the president of the Republic of Uzbekistan, had this to say about our people: "… our purpose is to create a worthy life for millions of families living in Uzbekistan. Only that will eventually provide power and riches for the state. This is and will be the most important factor defining in our policy" [1]. In 2012 at a cabinet session, the president of the republic once again emphasized, "At the center of our attention were questions of the further development of the social sphere, and a steady increase in the income and standard of living of the population of the country." The realization of strong measures in strong social policy runs like a red thread through all the years and stages of reform of our state.

The state is the subject of social policy, and the state defines social policy as activity focused on coordination of the efforts of all economic and administrative structures, and the entire population, with a view to the solution of social problems. The state plays a major role in social policy, carrying out economic development of the society as it is linked to the solution of social problems, such as employment, wages, social justice and social security. A market economy is not in a position to cope with their solution.

The primary goals which face to social policy are: (1) An increase in the well-being of people; (2) Realization of rules for social justice in society; and (3) Improvements in working conditions and quality of a life of the population.

Social policy acts not only in the social sphere, but also in the economic, political, cultural and spiritual spheres. Carrying out of state social policy is directed toward guaranteeing minimum income; maintenance and development of the abilities of members of a society, particularly abilities to work;

and maintenance of a comprehensive level of social services for members of society.

Social policy is implemented through legislative and statutory acts which define social policy actions, methods of social protection of the population, and also principles of creation of sources of financing through the process of distribution and redistribution of the national income. For example, at present something on the order of 17 % of the world gross national product (GNP) is spent in the social sphere. The developed countries spend on the order of 20 % of their GNP, whereas countries with a low level of income spend only 4 % of their GNP [3].

One of indicators used to define the effectiveness of social policy is the standard of living. Indicators of the standard of living are recommended by the United Nations. They include twelve groups of indicators: the demographic situation in the country—birth rate, death rate and other demographic characteristics; sanitary and hygienic conditions; food consumption; living conditions; formation and culture; working conditions and employment; income and living expenses; cost of living and consumer prices; vehicle ownership; the rest organization; social security; and human rights.

The listed indicators are considered core, and only an account including all of them can give a more or less exact representation of the standard of living in a given society. Experts of the "United Nations Development Programme" have developed generalized indicators of the standard of living, estimated as the average of three figures: gross national product by population (in US dollars on par with purchasing capacity of national currencies); life expectancy (a minimum of 25 years and a maximum of 85 years); and educational level (literacy of the adult population and the general coverage by training); these indicators are carried to the highest levels reached in the world. This combined indicator is called the "Index of Human Development". The United Nations publishes annual reports on human development where calculations are conducted calculations on index. In 2011 calculations were done on 187 countries. Table 1 contains data ranging a camp depending on the size of the change in the Index . The group of the countries included 47 countries with a very high level of human development, 47 countries with an average level, 46 countries with a low level. The Republic Uzbekistan has an indicator on the Index equal to 0.641 [3], which corresponds to the group of the countries with an average level. However, several factors will allow the republic to rise to new levels of social and economic development: rapid rate of economic growth (on the order of 8% annually); improving quality of medical care; an increase in average life expectancy; and nearly total literacy.

Social policy in Uzbekistan changed social priorities along with economic development. Here it

TABLE 1. The Groups of the Countries on an Index of Human Development

Ratings of the countries	Index of human development	Life expectancy at a birth (in years)	Average duration of training (in years)	Expected duration of training (in years)	Total national income (GNP) per capita (in US dollars for 2005)	Index of human development not connected with income
Very high level of human development	0.889	80.0	11.3	15.9	33,352	0.918
High level of human development	0.741	73.1	8.5	13.6	11,579	0.769
Average level of human development	0.630	69.7	6.3	11.2	5,276	0.658
Low level of human development	0.456	58.7	4.2	8.3	1,585	0.478

Source: http://hdr.undp.org

TABLE 2. **Dynamics of Expenses of the State Budget of the Republic of Uzbekistan**
(As a percentage of the general expenses)

Indicators	1999	2000	2010	2011 *	2012 *
Total expenses	100	100	100	100	100
Social sphere and social support for the population	31.5	45.5	59.0	58.7	60.0
▪ Education	17.9	23.2	33.3	-	-
▪ Public health services	10.0	8.7	12.8	-	-
Economics	40.7	9.2	11.3	-	-
Investments	16.6	20.4	6.4	-	-
Other spheres	11.2	27.9	23.3	-	-

Source: the Statistical collection. The basic tendencies and indicators of economic and social development of Republic Uzbekistan for years of independence (1990–2010) and the forecast for 2011–2015;
* According to a report by I.A. Karimov, President of Urzbekistan, at a session of the Cabinet of the Republic of Uzbekistan, "2012 becomes the year of a rising to a new level of development of our native land.".

is pertinent to indicate some stages in the change of directions of development of social policy.

In the first stage of economic reforms in the republic, state social policy was directed toward rendering assistance to the entire population. However such policy has appeared valid need of the population tolerant to degree. Further along, the process of deepening of economic reforms and development of the state social policy started to have an independent character, based on development of legislative and economic mechanisms to increase the standard of living. Since 1994 addressing the social protection of the population has been realized through mechanisms development after ascertaining the most vulnerable levels of population. The institute of local government of citizens (makhallya) began to play a considerable role, through which help was distributed to the most of people requiring it.

Now social policy is coordinated among structural divisions and is selective, regulating the most problematical questions of a social and economic character—processes of migration, the problem of formation, public health services, etc.

More than half of all annual expenses of the state budget in the Republic of Uzbekistan goes to the social sphere. Spending on social needs was on the order of 31.5% in 1999, 45.5% in 2000, and 58.7% in 2011 (shown in Table 2).

In 2012 more than 50% of the state budget was devoted to the development of education and public health services.

One of the main problems of social policy is the problem of carrying out distribution of income. Two indicators of differentiation of incomes use the factor of concentration of income, and are known as the Jeindex and Lorentz's curve. The factor size can vary from 0 (full equality) to 1 (incomes go to one person). The above value of the indicator, the боле incomes in a society are non-uniformly distributed. Figure 1 graphically shows the degree of inequality in incomes using Lorentz's curve.

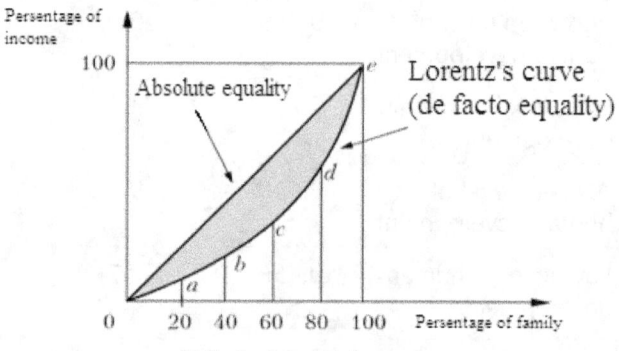

FIG. 1. Lorentz's curve.

This helps to define what share in the cumulative income is occupied by each group of the population, beginning with the poorest and finishing with the most well-off.

In the last ten years, our republic has increased the wages of the socially vulnerable segments of the population 8.1 times; this has been done by the use of tax privileges and general support. This has resulted in a decrease of 21.1 in the differentiation factor for incomes, decreasing it to 8.3. This is one of the lowest factors among the Commonwealth of Independent States as well as some of the economically developed countries [2].

Based on our findings, it can be noted that attention given in the Republic of Uzbekistan to the development and improvement of the social function of the state has paid off handsomely. It has resulted in maintenance of social protection, increases in the occupational level of the able-bodied population, an increase in the standard of living, and improvement in the well-being of the people.

References

1. I. A. Karimov Uzbekistan, on a threshold of achievement of independence. — Tashkent: «Uzbekistan», 2011.
2. I. A. Karimov becomes 2012 year of a raising on a new level of development of our native land,//the Report at session of the Cabinet of Republic Uzbekistan//the National word, on January, 20th 2012 г, № 14 (5404)
3. The report on human development 2011-http://hdr.undp.org/en/media/HDR_2011_RU_Complete.pdf
4. S. N. Ivashkovsky Macroeconomic: the Textbook. — 2 изд., испр., доп. — M: Business, 2002
5. The statistical collection. The basic tendencies and indicators of economic and social development of Republic Uzbekistan for years of independence (1990–2010) and the forecast for 2011–2015.

Strategy of Development of the Primary and Secondary Vocational Education in the Russian Federation

Sofia Morozova
Natalya Chernyshova

Russian State Agrarian University - Moscow Timiryazev Agricultural Academy, Moscow, Russia

For many years the absence of demand for young specialists on the labor market has been one of the most critical problems in the modern society. Besides, the deficit of the highly qualified workers and specialists of the middle rank is one of the most significant factors restraining the growth of all industries of the national economics.

Quoting Dmitry Livanov, Minister of Education and Science of the Russian Federation, today in the country «there are around five thousand educational institutions of the primary and secondary vocational education, where are around three million students are studying. In 170 Russian higher educational institutions the programs of primary vocational education (PVE) and secondary vocational education (SVE) are realized, and this adds another 200 000 students. To compare: today the enrollment of students of the higher educational institutions in the Russian Federation is six millions. That is, the PVE an SVE system is in general comparable with the higher education system by scale of its activity». [3]

According to the data of the Federal State Statistics Service, the number of students, admisson for studies according to the PVE and SVE programs and graduation of specialists with secondary vocational education classified by the sources of education financing were changing during the last years as shown below.

TABLE 1. Major indicators of SVE, thous. students.

	2000	2005	2010	2011	2012
Number of students (at the beginning of the academic year) - total	2308,6	2473,0	2026,8	1984,0	2087,1
Including students studying at the expense of:					
Federal budget	1145,4	1036,7	889,8	874,4	359,0
Budgets of the subjects of the Russian Federation	385,5	531,8	590,2	614,2	1162,1
Local budgets	59,3	35,3	5,7	3,3	3,1

TABLE 1 (*continued*)

With complete reimbursement of the cost of education	718,5	869,1	541,2	492,1	562,9
Number of students admitted - total	**842,4**	**810,9**	**671,8**	**628,8**	**656,2**
Including students studying at the expense of:					
Federal budget	377,4	329,7	291,2	280,4	114,8
Budgets of the subjects of the Russian Federation	119,7	159,9	196,5	198,5	365,9
Local budgets	19,4	11,0	2,1	1,0	1,2
With complete reimbursement of the costs of education	325,9	310,4	182,1	148,9	174,3
Graduation of specialists - total	**567,7**	**651,4**	**535,7**	**484,2**	**486,3**
Including students studying at the expense of:					
Federal budget	291,2	261,6	213,0	194,3	83,5
Budgets of the subjects of the Russian Federation	106,5	129,2	142,7	135,6	234,8
Local budgets	16,5	8,5	1,0	0,9	0,7
With complete reimbursement of the costs of education	153,5	252,1	178,9	153,3	167,3

The role of regional budgets in the SVE system is becoming more important, what is confirmed by the actual policy of reducing the financing of training in SVE institutions from the federal budget. Annually almost in all national higher educational institutions the number of state-funded places is cut off, while the state educational policy is aimed at their increase in the PVE and SVE institutions at the expense of the regional budgets.

In order to attract applicants, by the Federal Law «On Education in the Russian Federation» dd. the 29th of December 2012 № 273-ФЗ, and effective since the 1st of January 2013 and also by the Decree of the Ministry of Education and Science of the Russian Federation dd. the 28th of January 2013 № 50 «On the procedure of admission of citizens for education under educational programs of the secondary vocational education for 2013/14» the entrance tests to colleges, technical schools, specialized schools were aborted, except those specialties where certain artistic skills, physical and (or) psychological capacities are required [1, 2]. The certificate of secondary education was sufficient for admission. However, the percentage of applicants did not change cardinally. For example, according to the data of the Federal State Statistics Service, in 2013 the number of vocational educational institutions training middle rank specialists was reduced by 9% and amounted to 2703 institutions. The number of vocational educational institutions was reduced both among the state, municipal and private vocational educational institutions. Compared with 2012 the number of state and municipal vocational educational institutions was reduced by 8,7% and made 2488 institutions; the number of private vocational educational institutions was reduced by 16% and made 215 institutions. In 2013 (compared with 2012) the number of applicants admitted into vocational educational institutions training middle rank specialists was increased by 3,1%. Also in 2013 the tendency of redistribution of admission to vocational educational institutions training middle rank specialists in favour of the private sector was continued. Admission to private vocational educational institutions in 2013 was increased by 31,1%; at that, the programs of training middle rank specialists are realized by several private higher educational institutions, in 2013 the number of such institutions reached 112. [5]

The admission quotas by directions of training were changed as follows.

TABLE 2. Changes in admission quotas by professions and specialties of SVE with orientation at regional labor markets [6]

УГС		НПО			СПО		
		КЦП 2012 (человек)	КЦП 2013 (человек)	Изменение, в %	КЦП 2012 (человек)	КЦП 2013 (человек)	Изменение, в %
Гуманитарные науки	30000	249	100	-59,8	4210	3989	-5,3
Экономика и управление	80000	25	0	-100	14435	12038	-16,6
Сфера обслуживания	100000	956	1760	84,1	6680	5185	-22,4
Транспортные средства	190000	8968	7311	-18,5	12921	11394	-11,8
Технология продовольственных продуктов и потребительских товаров	260000	20552	22098	7,5	3811	3627	-4,8
Архитектура и строительство	270000	18730	17577	-6,2	7103	6968	-1,9

УГС – groups of specialties
НПО – PVE
СПО – SVE
КЦП – admission quotas
Человек – students
Изменение, в – change in %
Гуманитарные науки – Arts
Экономика и управление – Economics and management
Сфера обслуживания – Services
Транспортные средства – Transport facilities
Технология продовольственных товаров и потребительских товаров – Food and consumables technology
Архитектура и строительство – Architecture and construction

For example, admission quotas of SVE institutions by all directions of training were reduced, and to a greater degree they were reduced in the groups of specialties «Economics and management», «Services», «Transport facilities». At the same time, the admission quotas of PVE institutions were increased in the groups of specialties «Services» and «Food and consumables technology»; in the groups of specialties «Economics and management» and «Arts» the sharp drop of admission quotas is observed.

In general the dynamics of admission quotas in PVE and SVE is the following.

TABLE 3. Changes in admission quotas

Уровень образования	КЦП 2012 года (человек)	Предложения на 2013 год (человек)	Изменения (%)
НПО	96156	97332	+1,2
СПО	100482	97621	-2,9

Уровень образования – educational level
КЦП 2012 года (человек) – admission quotas of 2012 (students)
Предложения на 2013 год (человек) – supply for 2013 (students)
Изменения – changes
НПО – PVE
СПО - SVE

Important and probably major reasons for such changes are discrepancies between the state policy in the education sector, labor market demands and requirements of employers. Hence, one more reason of unattractiveness of PVE and SVE programs among the youth is revealed, this is the youth's awareness of necessity to get higher education. Graduates from SVE and PVE in the economics and management sector are oriented to further continue their training in a higher educational institution.

That is why «it is so important that the attractiveness of the SVE institutions and programs for the youth is stimulated. The efforts should be made so that the PVE and SVE organizations are addressed by not those applicants who failed to be admitted to higher educational institutions, but by those young people oriented to make successful carrier in the production sector». [2]

However, notwithstanding the long-term character of the outlined problems in the PVE sector, the state policy targeted to solve these problems is still inefficient and the corresponding legislation remains contradictory. For example, by the Federal Law «On Education» the PVE is completely abolished as an educational level (Article 10) and is equaled to the secondary vocational education under the programs of training of qualified employees (office workers) (Article 108). However, no meaningful changes were made in the training programs, while the government employees are still thinking of PVE as a separate educational level.

The discrepancies are observed also in the speeches of the Minister of Education and Science himself. For example, Dmitry Livanov states that «students of PVE and SVE system are often treated as losers who failed to be admitted to higher educational institutions, as outsiders in life» and calls for changing this situation. But at the same time he declares: «We understand very well that very often the PVE and SVE are being applied by the children from socially deprived families». O. N. Smolin, the First deputy chairman of the State Duma Committee on Education and Science, states that «there is a constant pressure to increase financing of national education by around two times», while financing of SVE has been reduced by almost 5 times for the last 3 years.

Besides, at present all SVE institutions are transferred to the regional level and thus the country is faced with another most important target – to increase the investment attractiveness of such education.

To solve this problem, for several years the government has been supporting the PVE and SVE institutions which implement innovational programs of training the labor force and specialists for the industrial sector. According to the information from the Department of State Policy in the Sector of Training the Labor Force and Further Vocational Education, in 2013 in the Russian Federation 4 444 state educational institutions (1 719 – PVE and 2 725 – SVE) and 200 state educational institutions of higher vocational education (HVE) realizing the PVE and SVE programs, were functioning; moreover, 256 private educational institutions of SVE with total enrolment of 102,7 thousand students and 16 structural departments of the higher educational institutions realizing the SVE programs with total enrolment of 3,3 thousand students, were functioning.

In 2007-2009 period 341 state educational institutions of PVE and SVE in 64 subjects of the Russian Federation have received the government support in the amount of 8,8 billion Russian Roubles, and another 8,7 billion Russian Roubles (co-financing) from the subjects of the Russian Federation and employers. In 2011-2013 period 30 subjects of the Russian Federation were backed up by the government support of 1,8 billion Russian Roubles (co-financing of 10,3 billion Russian Roubles) and realized the programs of modernization of regional systems of vocational education. As it can be noted, the budget financing from the Federal budget during these years has been reduced by almost 5 times.

In the period from the 4th till 7th of October 2013 during the Saint-Petersburg international forum of vocational education the All Russian conference on «Problems and perspectives of the secondary vocational education in the Russian Federation» took place and in the course of this conference the results of the federal competition of «100 best higher educational institutions of the Russian Federation» of 2013 were finalized, among which are the following institutions:

- State funded educational institution of secondary vocational education of the Ministry of Education «Krasnogorsk State College»
- State funded educational institution of secondary vocational education «Saint Petersburg Technical College of Management and Commerce»
- State autonomous educational institution of secondary vocational education of Moscow oblast «Gubernskiy Professional College»
- State funded educational institution of secondary vocational education «Nevinnomyssk chemical college»
- State autonomous educational institution of secondary vocational education «Bryansk Basic Medical College»
- State funded educational institution of secondary vocational education «Ufa Auto Transport College»
- Oblast administration state educational institution of primary vocational education «Vocational school № 2 in Voronezh»
- State funded educational institution of secondary vocational education «Moscow oblast college of information technologies, economics and management» of Moscow oblast
- State funded educational institution of secondary vocational education «Nizhniy Tagil State Teacher's Training College № 2»

- State Educational Government Financed Institution of secondary vocational education of «Gorno-Altai state polytechnic college» of the Republic of Altai.

Besides, in the period from the 15th of October till the 15th of November the Ministry of Education and Science of the Russian Federation has made the competitive selection of regional programs of development of vocational education for 2014 and 2015 in order to grant subsidies to the budgets of subjects of the Russian Federation and thus support measures of the Federal target program of education development in 2011-2015, and by the results of this selection 45 subjects were selected. In the top ten the Saratov, Kurgan regions, Khanty-Mansiisk autonomous district-Yugra, the Republic North Ossetia-Alania, the Republic Bashkortostan, Chuvash Republic, Volgograd region, Astrakhan and Tula regions, Khabarovsk Territory. Thus, the granted funds must be used for the enhancement of material and technical resources of colleges, training of specialists in demand, employment of graduates, implementation of new educational standards.

The experiment of implementing the Practical Bachelor's program, in which almost all federal districts participated, excluding North Caucasus District, has become an innovation.

TABLE 4. Distribution of participants of the experiment by federal districts

Federal districts	Number of Federal State Educational Institutions of higher education by SVE programs	Number of Federal State Educational Institutions of SVE
Central	2	12
Northwestern	1	5
Privolzhsky	1	6
Ural	1	2
Siberian	-	3
Far Eastern	1	1
Southern	-	4
North Caucasus	-	-
Total	6	33

The admission to the following specialties was carried out: metallurgy, engineering and material processing, computer sciences and computing technics, economics and management, education and pedagogics, energetics, power engineering and electric engineering, instrument engineering and optical engineering, aviation and aerospace technics, chemical- and bio-technologies, automatics and controls, electronic engineering, radiotechnics and communication, arts. And though so far it is too early to speak about the results of this experiment, however its initial results have already proved the unprofitability of such innovations as evidenced by the reduction of the number of students training under the programs on the basis of SVE, from 1145 students in 2010 to 672 students in 2012, and the major reason for it is absence of draft determent for young men. Besides, the deviation between practice and legislation is again observed: in the Federal Law «On Legislation in the Russian Federation» the Practical Bachelor's program is absent as a separate educational level.

In order to improve the quality of education the state policy was also targeted at adoption of the Federal State Educational Standard (FSES) of the 3rd generation by institutions of PVE and SVE. According to the analytical review of the results of the frontal monitoring in the institutions of PVE and SVE conducted in 2011-2012 academic year with educational institutions of all subjects of the Russian Federation participating, the adoption of the FSES of the 3rd generation was made by more than 90% of respondents. However, the problems encountered by the educational institutions while adopting the new standards turned out to be much more than it was anticipated.

The monitoring has revealed the following problems:

- absence or insufficient number of textbooks, methodological recommendations, modern equipment for training process does not favour the activization of professional activities of the pedagogical community and this can lead to the risk of information-technical failure of the training process and training of graduates according to

the requirements of the labor market and employers;
- low level of methodical provision of FSES fulfilment on the federal level, development of training programs, methodical and didactic provision of every educational institution systematically without any schedule, with the subsequent risk of destruction of the common national education space and limitation of the occupational mobility of graduates and the risk of «collapse» of the overall process of vocational education;
- rigidity of education system that results in the constant system-wide risk of training the personnel not complying with the demands of the labor market;
- professional competence of the staff of educational institutions who realize the personnel training is often not sufficient for fulfilment of their mission. In the situation when the single information database of the training program is absent, the work of teachers is «pushed back» to the development of the programs in the prejudice of development of the methodological support of the training program. The teachers are noted to have low motivation that leads to the risk that complying with the requirements of FSES will be treated as «optional», the same concerns the retrieving of truthful information about the results of FSES fulfilment;
- problem with expertise of effectiveness and quality of the educational process, mainly, who are the experts, how they are selected, what are the criteria of their selection, is the disbursement suitable for the specialty or professional profile? Lack of systematic multistage monitoring and feedback with employers leads to the risk of getting unreliable information about requirements in competence and qualifications. This is a direct challenge to the labor market because there is the risk that the number of graduates who do not correspond to the demands of employers will be increasing;
- not solved questions of organization of probations of teachers and masters of vocational training, mainly, what will be the subject of probation: skill upgrading of teachers, acquisition of the contents of a certain professional module, who will conduct the teacher's or a master's probation. Also there is a question what the legality of such form of training is from the point of view of its recognition in the course of the further accreditation of the teaching stuff and what document the employee will receive after his probation is finished. Hence there are risks of low effectiveness of probations;
- no elaborated system of target priorities in personnel training and, as the result, the risk of inferior quality of education. [4]

One more important step in this field has become the development of the «Strategy of development of the system of regular labor force training and formation of the practical qualifications for the period till 2020», the purpose of which is to create an up-to-date system of regular labor force training and forming the practical qualifications in the Russian Federation, which is able to:

- provide the quality and effective training of qualified employees (office workers) and specialists of the middle rank according to demands of economy and society,
- react flexibly to social-economic changes,
- offer wide possibilities for different social groups and communities in the acquisition of necessary skills and practical qualifications during all the working carrier. [7]

Hence, during the last years in the sector of PVE and SVE a sufficient number of programs was developed and realized, and they were very massively financed. However, for the time being their results introduced almost no practical changes in the situation on the labor market.

According to the forecasts of experts made till 2020, the gap between the number of graduates from PVE and SVE and the demands of the labor market will be kept for a long time and be more than 100%.

FIG. 1. Correlation between the number of PVE and SVE graduates and demand

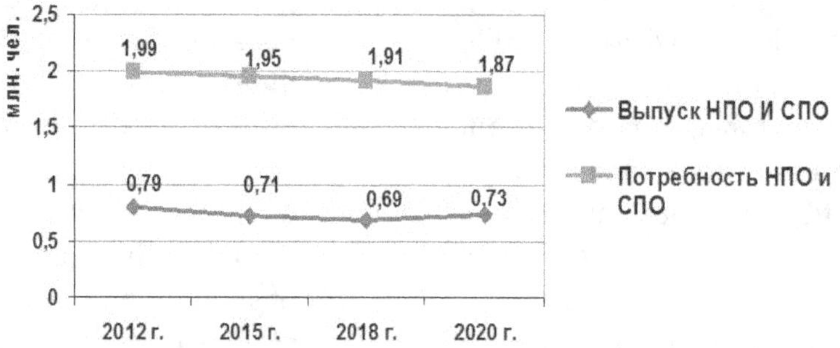

млн. чел. – million students
Выпуск НПО и СПО – number of PVE and SVE graduates
Потребность НПО и СПО – demand from PVE and SVE

At the same time, by forecasts of specialists made for 2014-2017, the following three specialties will become of the highest demand:

- engineers;
- IT-specialists and software developers, graphic interface designers;
- specialists in the service sector, for example, logistics specialists, ecologists, chemists, marketing specialists.

At that, it must be noted that every year the specialists are imposed with higher requirements. Regardless their field of activity, they must be ready to fulfil simultaneously several functions within the scope of their specialty at a time, and consequently, both the quality of education and the wages of specialists should be improved in direct correlation with the requirements. According to the data of the Federal State Statistics Service the average monthly nominal accrued wages of the employees of organizations by the type of economic activity in the Russian Federation during 2009-2012 were the following.

TABLE 5. Wages of employees of organizations by the type of economic activity, in Russian Roubles.

	2009	2010	2011	2012
Total economy	18637,5	20952,2	23369,2	26628,9
Agriculture, hunting and forestry	9619,2	10668,1	12464,0	14129,4
Fisheries, fish-culture	22913,5	23781,9	25939,9	29201,4
Mining	35363,4	39895,0	45132,0	50400,6
Manufacturing	16583,1	19078,0	21780,8	24511,7
Production and distribution of electric energy, gas and water	21554,2	24156,4	26965,5	29437,1
Construction	18122,2	21171,7	23682,0	25950,6
Wholesale and retail; repairs of auto transport vehicles, motor-cycles, household goods and personal appliances	15958,6	18405,9	19613,2	21633,8
Hotels and restaurants	12469,6	13465,8	14692,5	16631,1

TABLE 5 (continued)

Transport and communications	22400,5	25589,9	28608,5	31444,1
Financial activities	42372,9	50120,0	55788,9	58999,2
Real estate operations, rent and servicing	22609,7	25623,4	28239,3	30925,8
Public administration and military security provision; social insurance	23960,0	25120,8	27755,5	35701,4
Education	13293,6	14075,2	15809,1	18995,3
Public health and social maintenance	14819,5	15723,8	17544,5	20640,7
Other municipal, social and private services	15070,0	16371,4	18200,3	20984,5

As it can be seen from the table, the employees in the «Agriculture, hunting and forestry» sector earn the lowest wages, while the highest wages are earned in the «Financial activities» sector. These indicators explain why applicants choose a certain direction of training.

Thus, in order to solve all the above-mentioned problems and improve the quality of education in the institutions of the PVE and SVE with regards to the situation on the labor market, it is necessary to develop the complex of following measures:

- to restore the PVE as the separate educational level;
- to improve the legislation and eliminate its contradictions as regards the institutions of the PVE and SVE;
- to enhance prestige of the institutions of the PVE and SVE both among the youth and in the society in general;
- to increase scholarships for students;
- to secure the necessary financing from regional and local budgets as well as from the federal budget;
- to increase the wages of teachers of the PVE and SVE institutions;
- to make rejuvenation of teaching staff;
- to improve educational-material resources;
- to make the vocation education more attractive for investments;
- to equip the institutions technically;
- to eliminate the above-mentioned problems while the institutions of the PVE and SVE are switching to the FSES of the 3rd generation;
- to increase the wages for the middle rank employees and specialists;
- to develop public-private partnership;
- to provide interrelations between the PVE and SVE institutions and employers and develop employer-sponsored education.

Hence, nowadays the PVE and SVE are a significant branch in the system of education. Without a wise policy, improvements in legislation in this branch its successful functioning is simply impossible. Quoting Dmitry Livanov «only by consolidating our efforts, we together can form up and put into life the new modern system of training of middle rank employees and specialists in the Russian Federation that we need so much now» [3].

References

1. Decree of the Ministry of Education and Science of the Russian Federation dd. 28th of January 2013 № 50 «On the Procedure of the enrollment of citizens for education under educational programs of the secondary vocational education for 2013/14»
2. Federal Law «On Education in the Russian Federation» dd. 29th of December 2012 № 273-ФЗ
3. D.V. Livanov, Minister of Education and Science of the Russian Federation// VII Congress of directors of specialized secondary schools of the Russian Federation// Monthly theoretical and scientific-methodological magazine «Secondary Vocational Education», 8'2013
4. Project of the Federal Target Program of Development of Education «Analytical review of the results of the frontal monitoring in the institutions of PVE and SVE

conducted in 2011-2012 academic year»: http://www.myshared.ru/slide/273980/
5. http://www.gks.ru/
6. N. M. Zolotareva, Director of the Department of the state policy in the field of training of regular labor force and further vocational education «On planning of the volumes of training of labor force with vocational education (primary vocational education, secondary vocational education, further vocational education: http://www.marstu.net/Portals/Public/troubles/2013_0035.pdf
7. Strategy of development of the system of labor force training and formation of the practical qualifications in the Russian Federation till 2020

Capital Management Through the Lense of Economical Model of Timing Cycles

Andrey Nesterov

Russian Academy of State Service at the President of RF (Izhevsk branch), Izhevsk, Russia

Stock markets of different capitalist countries are nothing else but the blood system which delivers money (one of the major inventions of the mankind) to their destination. Also this is the place where the most ambitious and venturous people are fighting for their place under the sun.

Economic space does not always coincide with the rhythm of the physical world. All other social realities are interfering into this game. It should be pointed out that the world economy is spread through all the Earth, it represents the «market of all the world»[3]. Every global economy has a city pole which is a city right in the centre of ubiquitous elements securing its business activity: information, goods, capitals, credit, people, bills, business correspondence that flow in here and from here set off on the journey. Such cities nowadays are New York, London, Moscow, Tokyo (while their economical citadels are corresponding stock exchanges). It should be noted that the biggest role in this economic centre is played by the meaning of price. Price for any product, and in our case the product is shares, has its short-term or long-term character of the directed cycle. The tendency of price cycle is nearly invisible in this moment of time, in this very second, but it follows its own way always in one and the same direction, either increasing the price mass step by step up to a certain moment of time, or with the same insistency acts in the opposite direction starting the general decline of price, not unnoticed in the beginning, but fast, slow and inevitable[2]. Like any cycle, the cycle of share price growth or share price decline has its starting point, peak and final point and it is possible to determine these points with accuracy of one week or a day. By knowing the equation of growth and decline and the timing cycle, it can be defined with 99% probability when the price of a share or a currency or a raw material will grow or decline. Further on we shall show and prove this hypothesis on the real examples. It should be specified that the minute or the hour do not play any role in defining the timing cycle of share. A week-long statistical price analysis is necessary to determine the moment of growth of a certain financial instrument. In this case it is also important to define the sector of analysis to which a certain financial instrument is referred – shares from the public health sector, retail, banking sector, engineering sector. And only after that a certain share from a certain sector is chosen and the analysis of this share is made on the basis of the timing cycle and the equation of peak and bottom of the price range. Let us take two fresh examples from the stock market in the USA and the Russian Federation to make complete understanding of the above-mentioned information. Besides, it is important to note that the author is a trader and an asset manager with 6-year working experience on different stock markets. So, Moscow Central Stock Exchange, 2013, December (the month is important because for this sector in this case there appear certain economical indicators), in the equation of peak and bottom of the share price the situation occurred when the timing cycle has shown that the share of LSR at the price of 550 Russian Roubles per share is the temporary bottom. Knowing this, the author has purchased at 550-555 Russian Roubles per share. While creating the portfolio by these shares the scrupulous analysis of all actual market was made, and those shares appeared to be the best solution for purchase. We know that the markets, by their nature, cannot be always on the peak or bottom level and in the price flat. This is unnatural for the very sense of the market. So in this case the price of LSR shares after a certain decline (and the price bottom was easily defined for exact days) has grown up to 620 Russian Roubles from the bottom, or by 12-13%, and this means that the capital of the person who made such a decision, and in this case this was the author, was grown and showed the positive result. Based on the mathematics and knowledge about the timing cycles by economic sectors the theory of timing cycles was proved in practice.

Besides, some economists always had a very amazing point of view because they fix the movements of the market and describe these movement in the past time and are attentive towards many consequences of such movements, but they are always very surprised at the scale and infinite regularity of these movements, as if this cycle of growth or decline could not be predicted[3]. But this is not true.

Now some words about the stock market in the USA. Unlike the stock market in the Russian Federation, which is speculative due to many reasons such as liquidity, mentality, short historical period, the stock market in the USA is to a certain extent oriented at the investors and long-term increasing or decreasing movements. In this case the month, but not an hour or a day, is necessary to make the best investment decision. Summer is the time of vacations, but the author was relentlessly analysing the stock market in the USA and it became crystal clear that the steelmaking sector (all the market is calm) has shown the amazing vitality and liquidity. On the basis of the theory of timing cycles and the equation of price peak and bottom the share of the US steel was chosen and the entry point – ticker X – was defined. The price range for entry is 17.3 US dollars per share. The price at the time of writing this article in January 2014 is 28 US dollars per share. Pure mathematics and no miracles.

And every year there are many such examples on the global stock markets and also on the market of currencies and raw materials and the most important is to buy and sell on the basis of the timing cycles.

Most scientific theorists cannot explain why these upward and downward movements are developing and repeating. And nothing just happens. For example, our solar system is build just like the SYSTEM. If it was located plus minus two hundred kilometres to the left or to the right from the other planets, then our life would develop according to the completely different scenario. But this is just by the way. One major thing should be remembered: the most important criterion in the equation of the price peak and bottom on the stock market for any asset manager is TIME. And it does not matter into what time period (a month, a year or even several dozens years) you invest because it is possible to calculate the level of growth and the inherent pullback from the price peak, what makes the asset manager confident in the future. There is also an interesting fact that when the market is in the bottom after the sharp dropdown and when it is on the peak and everybody (mostly amateurs) is eager to buy, then the liquidity is increased by several times and thus the manager who is familiar and applying the theory of timing cycles gets the necessary amount of shares to fill in his portfolio and also has the liquidity for secure sales when the price will grow up.

Application of this equation is very interesting while detecting and overcoming the crisis periods. Crisis is the inherent part of the capitalist system. Let us remember the crises in 1763 and 1780 in the Netherlands. The foreboding of those crises was the excessive amount of «artificial money» on the market. And everything has its limits, and if the limit is exceeded than it is necessary to recede. In the situation of money redundancy those who possess the major pie of the capital start to buy shares and real assets, so prices grow up and up and the insane speculations begin, a bright example is the speculation with shares of the East India Company. And in the very end the peak of prices is outlined, then followed by a sharp dropdown, then comes panic and then again a dropdown. What is interesting is that if to take into account the timing cycle, it is impossible to get price-trapped if the situation develops in such a way. This is because the price peak always has characteristic economic features altogether with the timing features. Knowing these features, the purchase price will always be in the bottom point, while the sales price is close to the temporal peak. The crisis at Amsterdam Stock Exchange has begun on the 2nd of August 1763: «There is nothing to do in the stock exchange..... there is no exchange, there are no quotations, mistrust is everywhere»[3]. Does it remind you of anything? It is the same as the crisis in 2008, so everything is the same in all the times. Only TIME is unique. The crisis itself damages week firms from time to time, cleans the market from stock jobbers at one stroke and in general it is useful for the health of financial system, at least for those asset managers who have detected, based on the timing cycles, the point of price peak and who are working in the epicenter of this financial earthquake.

Sure, after the blizzard is over the market will grow up as fast as it dropped. And the manager should be in the centre of events so that to choose the correct shares for creating his investment portfolio in the right time and at the right price.

In one word, we can say that whatever the features of growth are, its movement has raised up the economy as a flood picking up the vessels dying in low waters; due to it the infinite and bonded consequence of inter-tied balances and disbalances have been born, it has delivered both easy and hard success, helped to avoid failures, invented profits [7]. It was a movement that again gave a new breath to the world after every slowdown or constraint. However, this movement, when it is downwards, is treated by many theoretic economists as the unexplainable «miracle» and is interpreted by them on the basis of numerous assumptions and hypotheses, though in fact just the TIME for growth has come and every dropdown cycle is changed by the growth cycle, as after the dawn always the sunset comes up. The market in total moves from one crisis to the next crisis, one engine is changes by the new engine, one source of growth is changed by another source, one pressure tool is changed by the other.

If to look at the market as the planetary capitalist system where the price of supply and demand is determined, such facts as mortgage crisis in the USA that was the cause of dropdown of all the global market in 2008, are of minor importance for the asset manager. Of course, he should take this into account while making decisions, but the major issue is not the reason why the market drops but when in started to drop. Because the causes of crises are different, if to take 1929, or 1998, or 2008 but the methodology and mechanisms of dropdown after the growth are always the same. Only masks of actors are changed but the scene is always the same.

There is an interesting question if the capitalism in its actual state will exist after 50-100 years. This question is very important for all the global financial system. Did crises of the twenty's century and of the twenty-first century broke the faith of mankind into the capitalist regime. Yes, there are unsatisfied people, there are people dreaming of socialism. But capitalism in general is one of the best systems that by our opinion is becoming more robust and stronger with every crisis. On both national and international level there is a redistribution, «brand new dealing out cards» - but in favour of the strongest players. The word «cards» in this context has figurative meaning because the author is not supporting those people who declare that the stock market is the casino where everything depends on the luck of players. Market is not a casino, but a place where the smart and patient people are becoming stronger and more wealthy while the «worker» himself is not a player but a professional in asset management.

The last word that we would like to say in this article is we described the practical experience of using the theory of timing cycles for the successful asset management. But this is only the top of the iceberg. By using this system and superseding the market and making profit which is higher than the market, one can manage the assets of the big company or a bank, the investment portfolio of the university by estimating the level of risk or profit, or manage even the whole state, in particular its investments sector. Money must work. And the profit must be invested into other projects, a part of it must be reserved in case of «the evil day», and another part of the profit must be invested into the real economy –into construction of infrastructure, social objects, and this is the correct asset management through the lense of the economic model of the timing cycles.

References

1. Belousov R.A. Economic History of the Russian Federation: XX Century. Book 1. M.,1999.
2. Brealey R. Principles of Corporate Finance. M.: Olymp-Biznes, 1997.
3. Braudel F. Le Temps du Monde. Progress. M. 1992.
4. James K. Galbraith. The Great Crash of 1929. Minsk: «Popurri», 2009.-256 p.
5. Jacobs N. Modern Capitalism and Eastern Asia. 1958.
6. W. D. Cohan. Money and Power. How Goldman Sachs Came to Rule the World. M.: Alpina Publisher, 2013 – 680 p.
7. International Financial Crisis: a New Stage of the Global Economic Development? M.: Institute of the Global Economy and International Relations, 1989.
8. Horne James C. Fundamentals of Financial Management. M.: Finances and statistics, 1996.
9. Ashton T.S. An Eeconomic History of England. The 18th Century. 1955.
10. Diamond S. The Reputation of the American Businessman.1955.
11. Hilferding R. Das Finanzkapital. 1910.

Theoretical Bases of the Aggregated Condition of Innovative Process

Evgenii Plakhin

Kursk State Agricultural Academy, Kursk, Russia

Abstract. *In this article the author suggests a typology of the general form of the innovation process. Conditions of efficiency of generation of innovations from the position of developers and managers of innovations are considered. System criteria of efficiency of innovative process are designated.*

Keywords: *structural innovation, niche innovation, local innovation, innovation process, technological structure, the effectiveness of innovation.*

Ensuring innovative production is a decisive factor in fostering competitiveness in the development of technological equipment for specific enterprises, and it is also a necessary condition of economic development of the country in which the enterprise is based. The technological structures comprising innovative components provide development of a made product, optimizing it to the general technological environment. This establishes criteria for satisfaction of requirements on the one hand, and forms the technological situation of development on the other. The impacts of an innovation on an economic system are subdivided into structural, niche and local; aggregated over time, they give the most concrete reflection of existing reality. For consideration of the properties and distinctive features of the specified categories, it is necessary to define structural, niche and local innovations. Structural innovations are systemic, transforming a significant number of niche processes and bringing them to new qualitative level of production efficiency in labor productivity indicators, processes that are the cornerstone of the technological component of production, and the quality and structure of satisfied requirements. Niche innovations fill a local vacuum in technological optimization, optimizing structural innovations for satisfaction of specific needs, and introducing technologies to increase the efficiency of existing processes. Local innovations are in essence innovations in thin control; their basic purpose is to increase the efficiency of niche innovations. Local innovations also carry out the important function of searching for the limit of application of a technology beyond which the efficiency of its application starts decreasing considerably. Structural innovations create a cascade effect of generation of niche and local innovations.

As an example illustrating our hypothesis, we can look at the invention and applications of the internal combustion engine (ICE). It is indisputable

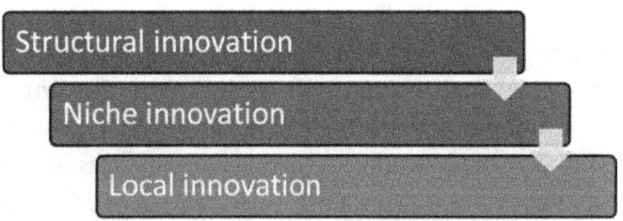

FIG. 1. General View of the Cascade Structure of the Aggregated Innovative Process

that the technology which is the cornerstone of the working principles of the ICE is a structural innovation in the system character of the engine. Niche innovation is application of ICEs for movement of payloads and people, that is, creation of automobiles; in this case the optimizing effect of the ICE technology is obvious. The local innovation in relation to ICE technology is shown in engine capacity change, decrease in fuel consumption, increase in efficiency, etc. That is, a number of small directed innovations that most fully satisfy requirements by means of thin control of technology. Here local innovations carry out the function of searching for of applicability of ICE. It is at present already obvious that peak of application of ICE technology has been reached, and the local innovations applied to this technology are directed at maintenance of the life cycle of what we consider to be the structural innovation. This is confirmed by that fact that the bulk of local innovations in ICE is directed at energy efficiency and the creation of hybrid propulsion systems in the automotive industry.

The typology of innovations given by the author is based on progress toward satisfaction of requirements by means of products and services based on technologies, which is the fundamental starting mechanism of the innovative process. Many other classifications of innovations exist; however, they are all reduced in the structure to the given classification and have a purely technical and specifying character.

From the standpoint of researchers, effective development of structural innovations is characterized by the following:

- An insignificant number of such developments is implemented, but sufficient manpower is used.
- Low level of transactional costs
- High level of a capital intensity

When developing a structural innovation, complex multilevel, branched organizational formations among researchers are not used. Rather, the structural innovations are created in the most compact organizational structures; the personnel possess all the necessary information and tools for carrying out research and development, or can receive them with the minimum transactional and financial costs.

From a position of managers, efficiency of development of a structural innovation is characterized by the following:

- The search for highly skilled personnel capable of coping with the objective
- Providing a minimum level of transactional costs, both in the organization of researchers, and between researchers and external factors
- Maintenance of an optimum level of a capital intensity of the research process

Managers are the membrane between outside effects and researchers. The level of transactional costs in the formation of development of a structural innovation between managers and researchers has to be a minimum.

During development of the conceptual basis of a structural innovation in the form of a model, a single copy or a small series of prototypes to carry out the desired function are created. The transition to the subsequent stage of the innovative cycle comes after establishing the working capacity and efficiency of the prototype.

In the course of integration of a structural innovation into a product, it is necessary for the company to provide effective functioning of engineering experts who deal with this problem for these purposes:

- Providing the engineering department with qualified personnel
- Low level of transactional costs
- Variability of application of a niche innovation with different types of products
- The maximum economic efficiency of a niche innovation in manufacturing lead time
- Aspiring to the maximum coverage of requirements

In making the decision to introduce structural innovations in an existing technological system, the structural innovation is transformed into a niche innovation in the course of optimization under satisfaction of specific needs. Creation of a niche innovation possesses the specifics which are determined by the technological basis of the production structure.

From a position of managers at the stage of creation of product efficiency, niche innovation is decided on according to the following factors:

- Search of engineering shots
- Low level of transactional costs, both in an engineering formation and in relation of the enterprise to the environment
- Ensuring economic efficiency of production and realization of the product of a niche innovation
- Providing economic conditions of stream innovative activity of engineering shots

For successful introduction of a niche innovation in production, managers need to provide engineers with a minimum level of transactional costs (as is also the case with structural innovations). In addition, the specifics of this stage create the need for innovative response to market conditions to achieve consumer satisfaction and create the best possible image given the conditions of the enterprise. A characteristic feature of niche innovations is that they have mass character and are available to a wide range of consumers, based on changes in the specific product. The impossibility of accounting for all features of consumption of a product, which is satisfaction of requirements for specific conditions, and maximum efficiency of production of a product creates a conflict situation. Formed a contradiction in total with inert nature of production containing a niche innovation becomes the main moving a factor appears local innovations.

Following the established tendency, we will define developers of local innovations as innovators. The main objective of innovators is to provide the maximum satisfaction of requirements by means of a niche innovation in a specific situation. The specific tasks are determined by the need to inject a local innovation into an existing product. During that process, shortcomings and directions of improvement for the product will come to light. To ensure efficiency in the introduction of a local innovation, it is necessary for the innovator to ensure the following:

- Qualification and knowledge of the specific consumption situation of the product which is the target of the local innovation
- Minimization of transactional costs for successful distribution of a local innovation if the innovator isn't the employee of the enterprise releasing the product
- No system effect of an introduced innovation
- Introduction of an innovation doesn't demand substantial reorganization of the product

Introduction of a product of a local innovation can take place in parallel with introduction of a niche innovation, and post factum at detection of features of use of a product. Owing to what thin control of a product containing a niche innovation comes to specific conditions of consumption. As a rule, the signal for development of a local innovation comes from consumers.

From a position of managers handling introduction of local innovations, the following actions are necessary:

- Tracking of the life cycle of the product and responses of users
- Search for third-party innovators and tracking of innovative initiatives by production of a product
- Decrease in transactional costs of introduction of local innovations in the organizational
- Ensuring economic efficiency of local innovations

Specialization of local innovations determines the characteristic tasks of a manager for this type of innovations. The tasks that enter here are: search and processing of user information on a product; the directions and methods of specialization of products; outsourcing of local innovations; and extension of the life cycle of a product . At this stage it is necessary for the manager to obtain sources of information to guide development, and adopt optimum administrative decisions concerning the development and use of local innovations. This requires carrying out appropriate market research to obtain information on consumption and learn of consumer opinion.

In all the cases we considered, both from the position of developers of innovations, and from

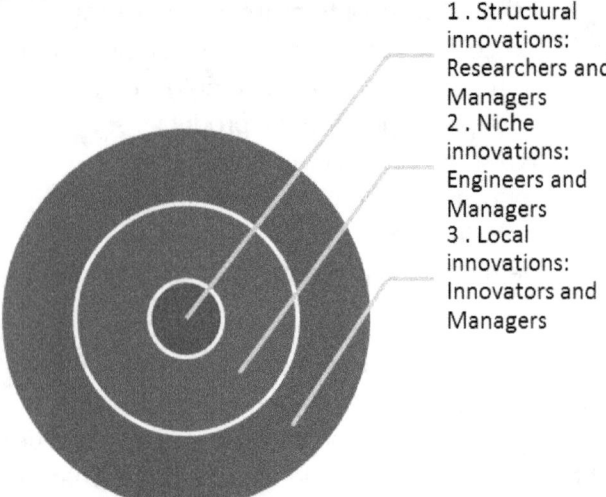

FIG. 2. Aggregated Organizational System of the Innovative Process.

1. Structural innovations: Researchers and Managers
2. Niche innovations: Engineers and Managers
3. Local innovations: Innovators and Managers

the position of managers at all levels, an important role is played by the need to decrease transactional costs. Providing a low level of transactional costs is the main objective of managers in the innovative process for effective functioning of developers of innovations and advances in innovations, from structural through niche to local. Transaction costs are the main factor in temporal inefficiency in the movement of an innovation to commercial realization. A high level of transactional costs is one of the major factors in loss of cost efficiencies in the course of production and product consumption.

The aggregated structure of the innovative process considered by us is presented in Figure 2.

In the first stage, a structural innovation which takes the central place in the innovative process reaches completion. In the second stage, niche innovations which optimize the structural innovation are determined and integrate the structural innovation into the product. The extreme importance of the second stage of the innovative process is due to the fact that failing to provide manufacturing lead time for the innovative process slows down the emergence of niche innovations. In this case even breakthrough structural innovations will reach consumers with considerable delays; the factor of a temporal inefficiency is shown here. In the third stage of the innovative process, local innovations, which on the one hand are proof of the success of applied technology, and on the other hand encompass movement toward the fullest satisfaction of requirements through thin control, are created. Local innovations form their own large market, which will be successful under a number of conditions, including openness of the platform of a product, the mass character of the product, and focus of the product on various tasks.

References

1. Coase, R. Y. «The Nature of the Firm». Economica. 4 (1937): 1, 386–405
2. Kline (1985). Research, Invention, Innovation and Production: Models and Reality, Report INN-1, March 1985, Mechanical Engineering Department, Stanford University.

Monetary Theory from Neoclassicists to Monetarism (Evolution Aspect)

Anna Sedova

Anna Ratzlaf

Orenburg State Pedagogical University, Orenburg, Russia

Abstract: *In this article the stages of evolution of monetary theory are considered. In particular, the analysis is given to main trends of monetary theory which were declared by the most significant schools and approaches in the economics: neoclassical school, Keynesian economics, modern monetarism.*

Keywords: *quantity theory of money (QTM); QTM equation; Keynesian economics; monetarism.*

The purpose of this article is to review the main trends in the money theory, in particular, to track the consistency of provisions of money theory which were declared by the main schools and approaches in the economics: neoclassical school, Keynesian economics, modern monetarism.

In modern economics there is a competition between several large trends of money theory. Monetarism and Keynesian economics have become the most popular.

We shall start our review from the traditional («old» or neoclassical) quantity theory of money (QTM), because all further QTM concepts were created on its basis.

«Old» QTM is normally associated with two formulas of money demand: Fisher equation (known as «Chicago equation») and Marshall equation («Cambridge equation»). These equations can be, in some sense, treated as equivalent if the difference between the consolidated income and volume of transactions is neglected. Indeed, the k coefficient in Marshall equation is the value reciprocal to the money velocity value in Fisher equation: $k = 1/V$ (2). However, the equivalence of «Cambridge» and «Chicago» equation should not be considered as absolute. Theoretically there is a qualitative difference between them. «Chicago equation» binds the money demand to current transactions, while by the «Cambridge equation» the money demand is considered as the demand for money balances, that is, as an asset of economic subjects which in fact must be opposed by other alternative assets. However, the «Cambridge equation» does not give direct answer to the question what determines the preference of one type of assets over the other type [1]

Certainly, the «old» quantity theory is not simplified to the unambiguous statement that the growth of money supply leads to the proportional increase of prices. In this theory several complicating factors are taken into consideration. However, finally after certain (not always short) period of time, according to QTM, the previous equilibrium in economy and the previous price level are restored.

However, if the quantity of money entering into the economy can provoke only a temporary deviation of the interest rate from its equilibrium, then what determines this equilibrium of interest rate according to the neoclassical theory? By opinion of neoclassical economists, the interest rate in economy does not depend on the quantity of money; it is determined by the supply and demand ratio of the loan capital. At that, in the neoclassical theory the demand for loan capital is equalled to the investments, while the supply is equalled to savings.

According to the neoclassical theory both savings and investments depend on the interest rate dynamics: as long as the saving interest rate is increasing, the investments are decreasing. Thus, the interest rate is determined by the demand and supply of the capital, but not of money. However, the increase of money quantity can invoke temporary shift of the interest rate from the level determined by the equilibrium between savings and investments. Keynes wrote in this respect that classical school in fact suggests two different contradicting theories: one theory binds the interest rate with the balance of savings and investments, the other theory binds it with supply and demand of money [1].

The contribution of Keynes into the economic theory is associated, in particular, with the revision of classical doctrine of interest and money. The level of savings, as Keynes supposed, depends first of all on the level of consolidated income and not on the interest rate. While the interest rate is the payment for sharing the liquidity, and the intensity of the liquidity preference depends on the demand and supply of money. Consequently, according to Keynes, the interest rate is the money phenomenon which is created on the market of money and alternative assets (bonds) and which is set to the real sector somewhat from the outside [2].

At the same time in Keynes system the interest plays critically important role in stimulation of investments on which the growth of consolidated income and employment depends. It is the interest rate correlated with the schedule of marginal efficiency of capital that determines the value of this marginal efficiency and, consequently, also determines the volume of investments in these conditions.

Thus, according to Keynes, the level of investments in the real sector is determined by the interest and consequently, the interest influences the effective demand, employment and production, however the interest rate is formed on the financial market.

Now let us dwell on the monetarism theory (which is the modern version of the quantity theory of money), the acknowledged leader of this theory is M. Friedman.

First of all it should be noted that M. Friedman considers monetarism exceptionally as the theory of money demand but not a price theory or a dynamic economics theory. Friedman underlines that in order to explain the price dynamics and consolidated income it is necessary to study several other factors besides changes of money quantity. The money demand is initially interpreted by Friedman in accordance with Keynes as the demand for a liquid asset which depends on the interest yield from other less liquid assets. However, the motives of money savings are treated by Friedman somewhat differently from Keynes. For Friedman the money is first of all the «luxury» and money savings are increasing as soon as income and wealth are altogether growing. Then, while in Keynes theory the bonds are the only alternative to money, in Friedman theory several income-earning assets (bonds, shares, real estate, physical capital) and also investments in the human capital are introduced as the arguments of the function of money demand [5].

However, Friedman studies the function of money demand on the large time scale. Based on the analysis of price dynamics and the quantity of money in the USA, he comes to conclusion that on the century scale the changes of interest yield from different alternative assets do not make influence on the money demand. Consequently, the only variable specifying the dynamics of money demand during such a long term is the consolidated income [5]. In fact, Friedman returns to the Cambridge equation of the quantity theory of money, however under reservations that k coefficient is increasing (in the long term) so far as the income is growing, while in the term of a business cycle totally different, sometimes divergent dependences are in effect. For example, within a business cycle the economic boom and income growth are accompanied by diminution of demand for securities and, vice versa, the depression and decrease of the consolidated income is accompanied by expansion of money demand etc.

Friedman assumes that on the basis of the century-long regularities of money demand it is possible to forecast the growth of this demand during the observed period and, consequently, to build up the central bank's politics of money supply in such a way that the balance of demand and supply is maintained and the price stability is kept to such an extent to which the prices depend on the quantity of money. However, in this case, as Friedman fairly notes, in their money demand the individuals are motivated not by the century-long tendencies but by their life experience and expected income. The income expected by individuals is relatively stable because individuals extrapolate to the future some average, stable income which they were receiving during several previous years. Therefore, the money supply should also be maintained on the stable average level and should not fluctuate following the fluctuations in the economic situation [1]. These fluctuations are quite difficult to predict and the reaction of money authorities to these fluctuations will be inevitably lagging behind and therefore will rather be harmful than useful. And vice versa, the balanced growth of money quantity according to the defined century-long trend and expected (average) income is able to make the economic system relatively stable. This is the main pattern of the monetarism approach.

Principles of Keynesian-neoclassical synthesis were born in the writings of J. R. Hicks and A. Pigou. Hicks has build up a model comprising the real sector where real savings and investments are made, and the monetary sector where money demand and money supply are confronted. The interest rates are determined in both above-mentioned sectors. In the real sector the interest rate depends on the conditions of equilibrium between savings and investments (Hicks has called it "investment rate"), in the monetary sector it is specified by the liquidity preference; it corresponds to the balance between the money demand and supply (Hicks has called it "money rate"). The main contribution made by Hicks in the monetary and interest theory is the statement that monetary interest rate is formed in the monetary sector and real sectors simultaneously.

In the monetary sector the major role in the interest dynamics is played by the principle of liquidity preference formulated by Keynes; the principle of equilibrium between savings and investments, developed by Marshall, still determines the interest rate in the real sector.

The contribution of Arthur Pigou into the neoclassical synthesis is first of all the explanation of so called "wealth effect". Wealth effect (Pigou effect) means that the money quantity is no longer considered as neutral with regards to the processes going on in the real economics, and this is true not only for the Keynesian approach, but also for the system of notions in the neoclassical theory. Thus, one of the main discrepancies between Keynesian and neoclassical theory is eliminated [1].

Therefore, as the conclusion we can note the following.

Quantity theory of money has a long history. In XVIII and in the beginning of XIX century it was developed in the writings of Hume and Ricardo, in the XX century it has been further developed in the works of I. Fisher and A. Marshall and has become the commonly recognized part of the neoclassical economic theory. In the 30-s of XX the quantity theory of money has become the subject to critics by J. M. Keynes who has formulated the alternative monetary and interest theory. Keynes was followed by J. R. Hicks and A. Pigou who laid the ground for the synthesis of Keynesian and neoclassical money theory (so called, «canonic Keynesian theory»). The modern variant of the Keynesian-neoclassical synthesis has been presented in the major work of D. Patinkin «Money, Interest, and Prices».

The alternative trend in economics is monetarism created in the works of M. Friedman where much is inherited from Keynesian theory, however, finally the conclusions of monetarism are close to the old quantity theory of money. In economics there is also a direction commonly known as «new Keynesian theory». In this theory the synthesis of Keynesian and neoclassical theory is impossible (the latter offers complete employment and accidentally achieved equilibrium) and the most attention is paid to the research of the non-equilibrium and involuntary unemployment.

Therefore, by our opinion, in economics there is hardly any single, «absolutely correct» trend that missions to reject all other directions. Every trend reflects a particular system of assumptions which simplifies the reality. In certain periods of time and in certain situations some particular system of assumptions can be more adequate than the other system; accordingly, the relevant economic theory can be more acceptable and «useful».

References

1. Blaug M. Economic Theory in Retrospect / M. Blaug; translated from English; M.: Delo (in Russian), 1994. 687 p.
2. Keynes J. M. The general Theory of Employment, Interest and Money / J. M. Keynes; translated from English; M.: Gelios ARB (in Russian), 2002. 352 p.
3. Rothbard M. Power and Market. Government and the Economy / M. Rothbard; translated from English; INFRA-M (in Russian), 1998. 452 p.
4. Williamson O. The Economic Institutions of Capitalism. Firms, Markets, Relational Contracting / O. Williamson. SPb.: Lenizdat (in Russian), 1996. 702 p.
5. Friedman M. If money could speak out / M. Friedman; translated from English; M.: Delo (in Russian), 1999. 160 p.
6. Harris L. Monetary Theory. Translated from English M.: Progress (in Russian), 1990. 445 p.

Development of the Concept of the Optimum Mechanism of Regional Government by Economy of Agro-Industrial Complex

Askar Sharipov
Gabit Ahmetzhanov

Zhetysu State University after Ilyas Zhansugurov, Taldykorgan, Kazakhstan

Abstract. In article the main mechanisms and problems of management are considered by agro-industrial complex and the concept of optimum control by economy of agro-industrial complex at the level of the region is developed. The developed concept improves realization of the mechanism of management by economy of agro-industrial complex and will increase efficiency of agro-industrial production of the country.

Keywords: *agro-industrial complex, optimization, management mechanism, efficiency, integration.*

Problem statement. In line with the market transformation of the agricultural sector the deep changes in its economic mechanism take place and the role of economic interests of all the subjects of market relations becomes more significant [2,257p.]. In this regards, in the reformation of the national agro-industrial complex an important role belongs to the government regulation of agriculture development for supporting the manufactures of agricultural commodities, providing the population of the Republic with the sufficient volumes of goods, securing the national safety.

The necessity to improve the management by the regional agro-industrial complex is explained by the fact that the government is responsible to its citizens for the creation of the normal conditions for living, for their procurement with the goods and other agricultural commodities. The lack of material and financial resources, low technical level and weak labor motivation, disruption of inter-industry relations, social backwardness of villages determine the necessity to provide better government support of this industry in the regions [9, 94p.].

At present Kazakhstan is ranked among 25 countries of the world which are major producers of agricultural raw commodities (corn and some other food products). Kazakhstan is ranked among the first ten largest producers of agricultural raw commodities in terms of the total territory, farming areas, cereals and potatoes crop acres. By experts' evaluations the resources of the republic allow producing 3 times more food products than it is consumed by its population. For the period of reforms since 1991 the 14,0 million hectares of tilled lands and 100 million hectares of pasture lands were withdrawn from agriculture. At that, the collapse of large commercial farms has led to the significant reduction of cattle stock, deterioration of material and technical resources, migration of population from villages and from country etc. Suffice to say that the cattle stock was reduced by 1,7 times, sheep and goat — by 2,3 times, swine — by 2,5 times, horse — by 1,3 times and poultry — by 2,1 times. Meat production was cut by 2 times. Reduction of production of the major types of agricultural products in 2010 compared to the 1990 has made: corn — by 1,8 times less, meat — by 1,8 times less, milk — by 1,1 times less, eggs — by 1,4 times less, wool — by 3,1 times less [3].

Relevance of this analysis is also connected with the insufficient development of the system of government by agro-industrial complex in many countries and absence of organization structures adequate to market relations and able to accumulate the material, labor and financial resources.

Mechanism of the government and market management of agricultural relations often in practice provides agricultural organizations with the choice of any form of production establishment, property, freedom in business activities, economic independence and free disposition of their agricultural commodities, and also of the income received after its distribution. In the conditions of the total crisis in the country, including the agricultural sector as well, where crisis in the beginning had drastic consequences not only for the agricultural industry, but for the country in general, the regulation of agricultural relations made by government has become a support to the deep economic, legal and social reforms [8,92p.].

All above-mentioned in general shows the objective necessity to make scientific research of the optimum mechanism of regional government by economy of agro–industrial complex.

The state mechanism of government by the economy of the agro-industrial complex has been in constant search since the years of transition to market relations. This theme is the subject of many research and development works, in particular what regards the theory and practice of the government management of the economy of agro-industrial complex in the period of transition to market.

The problem of government management of agricultural development is not new for the world and national science. Suffice to remember such names as A. Smith, D. Ricardo, J. Keynes, V. Varga etc.

Market cannot be created on its own, it is created by government on the organizational-legal basis being the most important element of its functioning. We can find the confirmation to this by J. Stiglitz: «...I am sure that in spite the fact that in the centre of success of our economy there is a market mechanism, the markets are by no means always organized on their own smoothly, and therefore they cannot solve all problems and they will always be in need of government as their most important partner» [6,328p].

The notion of the market economy as a liberal economic pattern is not absolute. «Liberalism has lighted up with more contrast the character of the market economy as a highly developed form of the anonymous exchange of goods and services», — as L. Erhard underlines [10,115p.].

G. Soros presents his point of view on the problem of market development and its basic goals: «...the structure itself cannot solve the questions of distribution in a fair way, because it takes the significant distribution of wealth for granted. The market does not reflect the interests of society in any way. The goal of the corporation is not to provide employment, they hire people (as few and as cheap as possible) only to make profit» [5,48c.]. This statement is the extra argument in favour of interference of government into the economy so that the received income is fairly distributed among the members involved into the production process. As the world experience has shown, the most effective form of the government support is the development and realization of special-purpose programs [2,257p.]. The importance of development and realization of programs so that to achieve stipulated targets was noted by P. Samuelson: «Namely government programs helped the USA to stay on the advanced positions in the scientific and technological sector» [4,54p.].

The questions of increasing the economic effectiveness of the agro-industrial complex have been and remain the priority for the economic theory in the scientific developments and their fulfilment in practice.

Quite a lot of attention to this problem is dedicated in the works of Russian academic economists — I. P. Dugin, L. I. Makarets, V. M. Bautin, F. K. Shakirov, G. V. Savitskaya, N. Y. Kovalenko, I. A. Minakova, A. V. Shpilko and others.

A significant contribution into the studies of these problems in the agricultural sector of economy was made by such national researchers as G. A. Kaliyev, T. I. Espolov, R. U. Kuvatov, J. J. Suleymenov, J. S. Sundetov, B. G. Zhunusov, K. R. Nurmaganbetov, K. K. Abuov, J. B. Balapanov, V.V., K. M. Belgibayev, H. V. Zharekeshev, S. Imangazhin, D. A. Kaldiyarov, A. K. Sharipov and others.

The role of the government in the transition to market relations is very well characterized by the academician G. Kaliyev: «Now applied in Kazakhstan, the principles of A. Smith that the «invisible hand of the market» will clear the air must be since a long time in the dustbin of economic history» [1,331c.].

The purpose of this research is investigation of **classical** methods of government and market **mechanism** of regulation of the economics of agro-industrial complex, development of the optimum mechanism of regional government by **agro-industrial complex** (AIC), improvement of the ways of effective implementation of innovation and investment decisions for the purpose of increasing the food security of the country.

This purpose is specified in the following **major targets**:

1. Investigation of the structure of government and market mechanism of management of economy of agro-industrial complex in the times of industrial-innovative development of the country;
2. Investigation of the potential of using the integration mechanisms, development of the market infrastructure of the agro-industrial complex;
3. Elaboration of methods stimulating the implementation of the scientific and technical achievements in the innovation and investment processes;
4. Improvement of the optimum mechanisms of the effective government and market management of the economy of the agro-industrial complex of the regions.

Main results of investigation.

1. Theory and global practice of the government regulation of development of agriculture shows that together with the market transformation of the agricultural sector its economic mechanism also undergoes deep changes and the role of economic interests of all subjects of market relations in the agriculture is becoming more important. That is why in the reforming of the agriculture an important role belongs to the government regulation of agriculture development for the purpose of supporting the manufacturers of agricultural commodities, providing the necessary volumes of food products to the people of the republic, guaranteeing the food security of the country. The necessity of the government regulation of the agriculture is caused by the following reasons:

- objective dependence of the agricultural production from the natural and climatic conditions;
- production;
- government's intentions to stimulate and achieve the food self-sufficiency and food security of the country;
- support of the income of rural manufacturers that supply the market with the products of general constant demand which do not sufficiently reimburse the expenses of manufacturers;
- creating parity relations between the agriculture and industries which belong within the sophisticated structure of the AIC.

Realization of radical social-economic reforms in the agricultural sector presupposes the acknowledgement of the arising problems from the new scientific point of view which accounts such aspects as interrelations between economy and environment, achievement of ecological security in rural areas. The process of transition to the market economy has given birth to the serious complications of ecology-economic and social nature that negatively influence the economy's growth rate, environment conditions, social standards of living and general perspectives of sustainable growth. Absence of real incentives for rationalization of land and water use and environment protection has resulted in absolutely low effectiveness of usage of land and water resources, environmental management and protection mechanism in the agriculture. Every year dozens thousands of hectares of agriculturally used areas, first of all, of irrigable lands are withdrawn from agriculture. degrading of agroecosystems, inferior soil quality, pollution of environment components is progressively increasing and the speed of this tendency is becoming a real threat.

At the same time, the transition to the new technology of industrial-innovation development based on the gradual refusal from environmentally harmful (environmentally intense) technogenic paradigm and transition to the model (strategy) of ecologically stable and economically balanced development is the vitally important stage of the social-economic progress. In general, aggravated ecological situation becomes the obstacle to the sustainable industrial-innovation development of the country and leads to the limitations in construction of manufacturing facilities and infrastructure, losing competitiveness of the national economy, strengthening of the negative influence of the ecologic factor on the conditions of agroecosystems and on public health, etc. Practiced agricultural production in general and exploitation of irrigable lands in particular is characterized by increasing tempo of involving the natural resources, depletion of reserves and their inferior quality. At the same time the science of environmental management and engineering is characterized by the increased interest to the issues of rational usage of natural resources and environment protection.

Maximum economic effect as the final goal of the agricultural production comes into conflict with the conditions of environment preservation, that is, it leads to the ecological disbalance or rupture of ecological sustainability.

The world experience and the further development of agriculture proves the necessity of the soonest transfer of the complex ameliorations on the new quality level where the optimum combination of economically effective measures complying to the principles of rational environmental management will be created and where the sustainable ecological development of agriculture will be maintained. Natural-anthropogenic equilibrium should be kept on the level which allows receiving the maximum economic effect at sustainable ecologic development of the agriculture.

Government funds should be assigned for financing the measures of improving the soil quality, land ameliorations; subsidizing the elite seed production, vegeculture, stock breeding and poultry; financing the capital investments into science and technics; creating the special state funds for financial support of agricultural producers; support of peasant (private) farms; reimbursing a part of expenses for purchasing the production infrastructure of the agro-industrial complex.

It is necessary to increase the number of government orders of agricultural products, raw materials and food products for securing the effectiveness of procurement for the needs of government. Such orders are also a form of guaranteed distribution of farm products, especially in the periods when the supply is much higher than the demand for agricultural products.

In the times when the republic enters into the Customs Union and into the World Trade Organization the following measures are necessary: sustainable development of the AIC sectors, maximum production of native-grown food products and consequently the decrease of the dependence of the country from the world import, resolving the social problem by employment of the rural population. Integration processes should be treated not just as the means of economic development of a region or an enterprise, but also as a factor of growth of their production and distribution base.

Development of the indicative and budgetary planning solves the problem of the government regulation of agriculture through adoption of the programme budget in which the distribution of state funds for the needs of agriculture is provided.

Crediting policy should be oriented at:

- making the loan interest less expensive;
- reimbursing the payments against the principal debt;
- allocating the concessional credit funds;
- deferring the taxes payable; establishing rural credit societies.

The number of leasing operations with participation of government should be increased because it should become a measure for government support of AIC. The second-tier banks should be the guarantors to the companies leasing the farm machinery and should take active part in these operations.

An important role belongs to the insurance measures for the purpose of insurance of the crops, stocks, poultry and investments into the agricultural production. The system of insurance of the agricultural sector is stimulated by sharing the insurance costs between the state budget on the one side and agricultural enterprises and farms on the other side. Premiums for insurance of crops paid at the expense of the own funds of the producers should be reimbursed and are referred to the self cost of agricultural products.

Conclusions.

1. In the reformation of agriculture the important role belongs to the government regulation of agriculture for the purpose of supporting the producers of agricultural products, providing the population with sufficient volume of products, guaranteeing the food security of the country.
2. The research specifies the measures of government support of agriculture, they are regulative in character and are fulfilled in the following directions:
   - development of targeted complex programs which are an important factor of influence of the government on the development of scientific-and-technological advance in agriculture, maintaining the national food security.
   - budgetary and credit policy oriented at regulation of the money and credit resources.
   - n government orders of agricultural products, raw materials and food products for securing the effectiveness of procurement for the needs of government. Such orders are also a form of guaranteed distribution of farm products, especially in the periods when the supply is much higher than the demand for agricultural products.
   - insurance measures for the purpose of insurance of the crops, stocks, poultry and investments into the agricultural production.
3. After analysis of the situation with government regulation of agriculture the following conclusions are made: radical social-economic reforms in agricultural sector should be made on the basis of interaction between economy and environment, environmental safety in rural areas should be achieved; the quality control of food products should be provided; measures restricting the dependence of the food products market from import should be realized; at least the minimum threshold volume of the national production of the basic food products should be provided.
4. After analysis of the situation with procurement of the population of the republic with food products in 2013 the following conclusions are made: the specific correction measures are necessary: guarantee of the economic growth in agriculture; realization of measures for decreasing the dependence of the food products market from import; creating the effective system of agricultural business through development of land relations and agricultural water use, improvements in the seed farming, restoration of agrochemical service, provision of farm machinery, development of farm machinery engineering, protection and plant quarantine and stock breeding (livestock breeding, veterinary medicine, beef and milk farming and dairy breeding). Significant target is the growth of competitiveness and volume of sales of agricultural products, development of processing industry and information-marketing system.

Perspectives of further developments: to solve the issues of government regulation of agriculture development it is suggested to proceed from the following major directions:

- combination of economic and social targets that provides solution of not only the problem of development of agricultural production and its maintenance sector, but also solution of social issues. In the Republic of Kazakhstan the most important issue is the increase of the level of material security and measures of raising the employment in agriculture considering that the unemployment there is rather high;
- economical incentive of production based on the revision of pricing and financial-crediting mechanism. This direction is grounded first of all on restoring the price parity, on protecting the interests of agricultural producers from

monopoly of other (first and third) AIC sectors which dictate economically unprofitable conditions of inter-sectoral exchange;
- program regulation based on the determination and realization of strategic directions of development of agricultural sectors. In Kazakhstan the above-mentioned refers to the functioning of wine-trade, fruit-and-vegetable cannery, cattle-breeding cooperatives;
- budgetary support of the top-priority goals of development of cooperation, creation of the multilayer agricultural sector with the reasonable correlation between different forms of agricultural businesses where the leading role belongs to the large agricultural enterprises that have preserved their production capacity (farm lands, industrial purpose projects, machinery, perennial plantings, productive livestock, skilled personnel) and to a greater extent correspond to the historically formed communal, collective mentality of the peasant population of the republic.
- government support of the top-priority directions of development of agricultural production that determine the scientific-and-technological advance (breeding and seed farming, livestock breeding, biotechnology).

References

1. Kaliev G. Selected works.- Vol.1. — Almaty; -2008.- 331p.;
2. Mishulina O. V.//Economical Efficiency of Agricultural Production: Theory, Methodology and Practice// Autoabstract in candidacy for the academic degree of Doctor of Economics. -2012-257p.;
3. Program of Development of Agro-industrial Complex of the Republic of Kazakhstan for 2013–2020. «Agrobusiness-2020» //agrosektor.kz›news/item…**programma…apk…2013**–2020;
4. Samuelson P. A. Economics. Vol.1.-M.: «Alfavit», -1993.- P. 54–55;
5. Soros D. Open Society. Reforming Global Capitalism.- M., -2001.-P.48–49;
6. Stiglitz D. The Roaring Nineties.- M.: Modern economy and law. -2005.-328–373p.;
7. Sustainable development strategies: a resource book/ compiled by Barry Dalal-Clayton and Stephen Bass of IIED. OECD & UNDP. First published in the UK and USA by Earthscan Publications Ltd., -2012. — 358 p.;
8. Syzdykov B. Economic methods of government regulation of the agricultural production of Kazakhstan. Published by «R-Com» LLP, -Almaty.-2011. –P.92–93;
9. Sustainable Solutions for Agro Processing Waste Management: An Overview\C. M. Ajila, Satinder K. Brar, M. Verma; Environmental Protection Strategies, -2012.- 94 p.;
10. Erhard L. Wealth for All.- M.: Beginning of progress, -1991.-115p.

Public Efficiency of the Energy Market: Concept and Measurement

Mikhail A. Simonov

Institute of Control Systems, Ekaterinburg, Russia

Abstract. *The electric power industry is a basic sphere and the core competitiveness of companies in developed and developing countries. The author is mainly interested in countries with the existing competitive market of energy generation. The main feature of the electricity industry in these countries is the tendency to monopolize local markets. In order to resolve this problem this article introduces a new concept of social efficiency of the energy market and presents a methodology which allows a balance to be established between the interests of owners of wholesale generating companies (WGC) in generating income and the interests of consumers of energy in reliable energy supply by finding a correlation between the stability of WGC and the level of competition in the market.*

Keywords: *social efficiency of the energy market, stability, wholesale generating companies (WGC), Herfindahl-Hirschman Index (HHI).*

The power industry is of strategic importance in terms of ensuring the competitiveness and economic security of any country with an advanced fuel and energy complex. [1] Therefore, the need to monitor and maintain the sustainable operation of generating enterprises, networking, sales and service companies should be a priority not only for their management, but also for the state. Forming local markets in certain regions or united energy systems (ECO), large wholesale generating companies seek to gradually reduce the level of competition, bringing the market to a state of monopoly. However, this process undermines the sustainability of smaller companies, which can cause problems with the reliability of power supply of the consumers in the region.

In order to evaluate the effectiveness of the market from the point of a compromise between the level of competition in the wholesale market and the stability of the generating company the author introduced the concept of "social efficiency of the energy market," which is determined by the ability to contribute to the solving of the following tasks:

1) involving a wide range of investors to the industry in order to fund advanced energy technologies;
2) provision of reliable and quality parameters of energy supply;
3) curbing the price increase;
4) introduction of advanced energy technologies in reconstruction or construction of new facilities in order to improve the reliability of power supply.

Public market efficiency is determined by the intensity of competition in terms of the impact on costs and tariffs, as well as reliability of supply based on the introduction of advanced energy technologies.

For the purposes of the research of the market of energy capacities and electricity supply in terms of their concentration in the electricity companies we considered it appropriate to apply four following indicators:

1) *The level of dominance of the four leading energy companies in the balance of power generating capacity* of a particular territory. The formula of calculation: UrDm = $(N_1 + N_2 + N_3 + N_4)$ / Nreg, where N_1, N_2, N_3, N_4 – the capacities of four regional power generation companies (while $N_1 > N_2 > N_3 > N_4 > N_5 > ...$), Nreg is the total of generation in the region.

2) *The level of dominance of the four leading energy companies in the balance of power production* in the specific area (region). The formula of calculation: UrDee = $(E_1 + E_2 + E_3 + E_4)$ / Ereg, where E_1, E_2, E_3, E_4 represent the production of electricity by the four regional generation companies (while $E_1 > E_2 > E_3 > E_4 > E_5 > ...$), Ereg is the total production capacity of electricity generation across the whole region.

3) *Energy concentration index in relation to the installed capacity* of the generation companies (eHHm). The formula of calculation: eHHm = $(N_1 / Nreg)^2 + (N_2 / Nreg)^2 + (N_3 / Nreg)^2 + ...$. The index ranges from 10,000 (monopoly) to 0 (infinite number of participants). For electricity, by analogy with the countries with developed market economies, the optimal energy concentration index is within the range of from 1000 to 2000. If eHHm < 1000 the influence of monopolies will be observed, if eHHm > 2000 the power capacity will be unduly dispersed which will not allow an active policy of updating and building generation capacity to be conducted.

4) *Energy concentration index in relation to the volume of production and sales* of electricity (eHHee). The formula of calculation: eHHee == $(E_1 / Ereg)^2 + (E_2 / Ereg)^2 + (E_3 / Ereg)^2 + ...$. The range of variation of eHHee is similar to the variation range of eHHm. Such indicators are widespread in the countries with traditionally developed market economies. The concentration index is known as the Herfindahl-Hirschman Index (HHI). Therefore, our task was to adapt the most effective indicators of assessing the level of concentration in terms of electricity and power market in Russia. That's why we added the Herfindahl-Hirschman Index to the similar adapted indicators for the merging of two generating companies, which is calculated as follows:

$$eHHI_2 = (D_1 + D_2)^2 - D_1^2 - D_2^2,$$

where D_1 is the share of the first company in the market; D_2 is the share of the second company. If the index $eNNI_2$ exceeds 100, the market becomes highly-concentrated. Under the circumstances where mergers are observed in the domestic electricity market, the use of this index is relevant.

Using the proposed metrics, we assessed the level of concentration in the generation market within the boundaries of the United Energy System of Ural. The United Energy System of Ural is located in the Ural and Volga federal districts of central Russia. Many large industrial companies operate in the region, they develop dynamically and form the basis of electricity consumption and power. In terms of installed generating capacity, the market concentration (index eHHm) as of 2011 was as following (Table 1).

Therefore, the total share of installed capacity of four of the nine energy companies operating in the region is about 56 %. According to this indicator, the concentration in the industry is weak. This is confirmed by the results of the calculation of the index eHHm, which is equal to 935.

Now let us consider the market share and the index eHHee for the United Energy System of Ural in the context of electricity generation (Table 2).

In this case, the total share of the top four companies in the market is 80 %. This indicates a high concentration of the market within the boundaries of the United Energy System of Ural. The existence of a high concentration is confirmed by the index value of eHHee which equals 1767. The difference in the values of the indices for the production and for the installed capacity is determined by the fact that four generation companies (leading by ratio) have a maximum number of hours of usage of the installed capacity.

Below we can see the results of calculations of the index $eHHI_2$ for different pairs of merging companies from a perspective of the index of the total installed capacity of companies, which make up the UES of Ural (Table 3). The highlighted combinations show the danger of merging for the concentration

TABLE 1. The calculation of the index of eHHm for the UES of Ural based on the data for 2011

Company	Installed capacity	Share	HHI
WGC - 1	7897	18,2%	
WGC - 2	5865	13,5%	
WGC - 3	882	2,0%	
WGC - 4	5400	12,5%	
WGC - 5	4982	11,5%	
RusHydro	1539	3,6%	
Energoatom Concern OJSC	600	1,4%	
TGC - 9	550	1,3%	
Bashkirenergo	4556	10,5%	
Total	43300	100,0%	
Index eHHm for the united energy system (UES) of Ural on the installed capacity =			935

of the companies in the market, the possibility of a monopoly creation.

Thus, the possibilities of the tested indicators include assessing the actual level of competition and measuring the effects of events that are planned for implementation in the industry.

Since the market of the United Energy System of Ural is highly concentrated as to the production, it is necessary to follow not only the actual company mergers (Table 4) inside it, which are unlikely to happen soon, but mainly to follow the transferring of the ownership (shares) over-the-counter. The

TABLE 2. Calculation of the index of eHHee for the UES of Ural based on the data for 2011

Company	Production, million kW/h	Share	HHI
WGC - 1	37677	21,7%	
WGC - 2	33128	19,1%	
WGC - 3	3920	2,3%	
WGC - 4	39165	22,6%	
WGC - 5	28355	16,4%	
RusHydro	4320	2,5%	
Energoatom Concern OJSC	4022	2,3%	
TGC - 9	2890	1,7%	
Bashkirenergo	19800	11,4%	
Total	173277	100,0%	
Index eHHee for the united energy system (UES) of Ural on the production =			1767

TABLE 3. The degree of influence of the mergers of power generating companies with the facilities within the boundaries of the UES of Ural on public market efficiency in the context of installed capacity

Installed capacity	WGC - 1	WGC - 2	WGC - 3	WGC - 4	WGC - 5	Energoatom Concern OJSC	TGC - 9	Bashkirenergo
WGC - 1		494,1	74,3	454,9	419,7	50,5	46,3	383,8
WGC - 2	494,1		55,2	337,8	311,7	37,5	34,4	285,0
WGC - 3	74,3	55,2		50,8	46,9	5,6	5,2	42,9
WGC - 4	454,9	337,8	50,8		287,0	34,6	31,7	262,4
WGC - 5	419,7	311,7	46,9	287,0		31,9	29,2	242,1
RusHydro	129,6	96,3	14,5	88,7	81,8	9,9	9,0	74,8
Energoatom Concern OJSC	50,5	37,5	5,6	34,6	31,9		3,5	29,2
TGC - 9	46,3	34,4	5,2	31,7	29,2	3,5		26,7
Bashkirenergo	383,8	285,0	42,9	262,4	242,1	29,2	26,7	

important point is the presence of various affiliates and structures, which are owned by wholesale generating companies and other energy companies that produce electricity.

Analysis of foreign experience in reforming the electricity industry and of the approaches to assessing the level of competition in some local markets allowed us to determine that the Russian power industry is developing in line with international experience. The data on the nature of the competitive environment in certain Russian electric energy systems sets us a mission to outline the range

TABLE 4. The degree of influence of the mergers of power generating companies with the facilities within the boundaries of the UES of Ural on public market efficiency in the context of electricity generation

Production	WGC - 1	WGC - 2	WGC - 3	WGC - 4	WGC - 5	Energoatom Concern OJSC	TGC - 9	Bashkirenergo
WGC – 1		831,4	98,4	982,9	711,6	100,9	72,5	496,9
WGC – 2	831,4		86,5	864,3	625,7	88,8	63,8	436,9
WGC – 3	98,4	86,5		102,3	74,0	10,5	7,5	51,7
WGC – 4	982,9	864,3	102,3		739,7	104,9	75,4	516,5
WGC – 5	711,6	625,7	74,0	739,7		76,0	54,6	374,0
RusHydro	108,4	95,3	11,3	112,7	81,6	11,6	8,3	57,0
Energoatom Concern OJSC	100,9	88,8	10,5	104,9	76,0		7,7	53,0
TGC - 9	72,5	63,8	7,5	75,4	54,6	7,7		38,1
Bashkirenergo	496,9	436,9	51,7	516,5	374,0	53,0	38,1	

FIG. 1. Rational share of WGC in the local electricity market

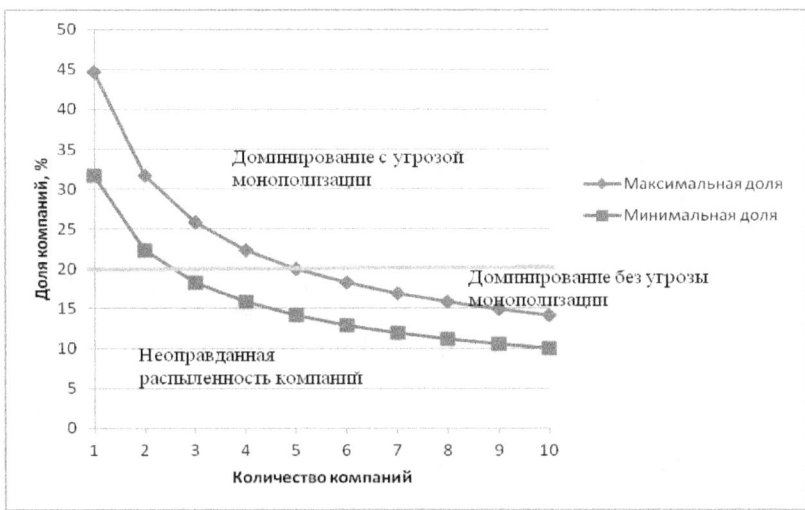

Доля компаний, %	The share of companies, %
Доминирование с угрозой монополизации	Domination with the threat of monopolization
Максимальная доля	Maximum share
Минимальная доля	Minimum share
Доминирование без угрозы монополизации	Domination without the threat of monopolization
Неоправданная распыленность компаний	Undue dispersal of companies
Количество компаний	Number of companies

The rational proportion for UES of Ural ranges from 10.5% to 15%.

of tools to assess the sustainability and to develop a mechanism for controlling the stability.

These indicators make up the basis for determining a rational fraction of the generating company in the local electricity market (Fig. 1). We calculate the value of rational WGC share in the local market from the equation for eHHm or eHHee under the assumption that the shares of all companies are equal:

$$Д = \sqrt{(T / n)}, \text{ where}$$

Д is the optimal share of the company in the market;

T is the value of index eHHm or eHHee;

n is the number of companies in the local market.

This range shows the "corridor" (shaded area in the figure), which contains values of rational shares of companies in the market with different numbers of them. Above the corridor we can see a zone where one or a few companies dominate with the threat of monopolization of the market. This is the zone of limited competition where the rates may be affected not only by the price factors. Below the "corridor" we can see the area of undue dispersal of the companies when their market share is low and on this reason the stability of each of them decreases. The distance from the "corridor" up to the level of 20 % (according to competition legislation) is characterized by the dominance with minimal threats for monopolization.

We should also note that the electric power industry is a leading infrastructural sector where it is highly necessary to eliminate the negative effects of companies' use of market dominance. Therefore, more hard criteria for dominance within this sector are set than in accordance with the legislation on the protection of competition or the criteria for other domestic industries. Thus, clause 3 of article 25 of the Law of Russian Federation "On Electric Power Industry" states that " the position of an economic entity (group of persons) is recognized to be the dominant, if the share of the installed capacity of the generating equipment or the share of electric power generation with the usage of the specified equipment within the zone of free flow exceeds 20 percent... ".

Thus, the more companies are located in the free-flow zone, the greater is the competition, although the size of each company in the market is becoming smaller, which leads to lowering the level of their stability. That's why, in order to achieve a compromise between the level of competition in the market and the stability of the generating companies working in it, we had formed a range of rational shares of companies in the market.

In conclusion, introducing the notion of social efficiency of the energy market and the method of finding a compromise between the interests of owners and consumers provided in the article, building a "corridor" of rational shares may be used by the companies to ensure the sustainability of its own functioning and development. It may also be used by the system operator to ensure the sustainability of the energy system as a whole and within local markets.

References:

1. Gitelman L.D., Ratnikov B.E. Energy Business: Textbook. - Moscow: Delo, 2008. – 600 pages (in Russian).

Axiological Approach to Application of Innovations in Work of the Small Commercial Enterprise

Maxim A. Titovets

Ural Institute of Business of a name of I. A. Ilyin, Ekaterinburg, Russia

Abstract. *Consideration in the theoretical plan of set of the concepts connected with an axiological (valuable) approach to conducting of business activity; judgment and disclosure of spiritual and moral aspect of application of innovations in work of the small commercial enterprise.*

Keywords: *axiological (valuable) approach; business activity; small business; subject of small business; innovation; spiritual and moral economy.*

Market transformations, their rates and success substantially depend on the solution of problems which are connected with development of the business activity defining in the modern world the most important economic processes and tendencies.

Transformation of business activity into a determinant of economic development directly depends on creation of conditions which allow in rather short terms to the resolute and initiative people possessing necessary qualities and abilities, to become civilized businessmen.

Establishment of a concrete circle of the aspects characterizing essence of any concept, is a starting point for a formulation of the purposes, structure and volume of further researches. Let's give particular attention to research of essence of the base concepts used in this article.

Axiological (valuable) approach.

It is considered that the concept «axiology» (from Greek. axia — value and logos — a word, idea) is entered into a scientific turn in 1902 by the French philosopher Lapi, and in 1908 it was used actively by German scientist E.Gartman.

In philosophical dictionaries the axiology is defined as a science about values. More developed definition is given in «The pedagogical dictionary» to G.Kadzhaspirova: an axiology — philosophical doctrine about material, cultural, cultural, moral and psychological values of the personality, collective, society, about their ratio with the world of realities, change of valuable and standard system in the course of historical development [9].

The person lives in a condition of a world outlook assessment of occurring events, it puts before himself tasks, makes decisions, realizes the purposes. Thus its relation to world around (to society, the nature, to himself) is connected with two approaches — practical and abstract and theoretical (informative). The link role between practical and informative approaches carries out an axiological (valuable) approach.

The axiological approach is aimed at identification put in a subject of knowledge of possibilities which would satisfy these or those material or spiritual needs both the most learning person, and other people. The knowledge undertakes from the point of view of its cultural parameters, from the point of view of possibility of knowledge to serve in this aspect to the person, his benefit, the benefit of people.

Ideas of an axiological approach.

Without man, and without the concept of value cannot exist, because it is a special type of man significance objects and phenomena. Values are second only, they are derived from the ratio of the world and man; values confirm the importance of having created man in the process of history. Values include only positively significant events and phenomena associated with social progress.

Principles of an axiological approach.

- equality of all philosophical views within uniform humanistic system of values (at preservation of a variety of their cultural and ethnic features);
- equivalence of traditions and creativity, recognition of need of studying and use of doctrines of the past and opening possibility in the present and the future;
- equality of people, a pragmatism instead of disputes on the bases of values; dialogue instead of indifference or denial of drags of the friend.

These principles allow to join in dialogue and in common to work to various sciences and currents, to look for optimum decisions.

Values — not only subjects, the phenomena and their properties which are necessary to people of a certain society and the individual as means of satisfaction of their requirements, but also idea and motivation as norm and an ideal [12].

The economic knowledge as a special case of scientific knowledge and human activity is through impregnated with values and without them it is impossible. Economic knowledge is always of value-oriented and this orientation defines the entire process of knowledge — from the choice of the object of knowledge, methods of knowledge to the practical use of the obtained results.

According to the standards accepted in many countries, business activity is the human activity, directed on the organization and implementation of important and difficult projects. It associates with attempts to make something new or to improve something already existing. The main role in this activity belongs to businessmen who reduce together money, material resources, labor, as a result create new business and operate it [5].

Business activity — initiative, independent, carried out on its own behalf, on the risk, under the property responsibility activity of citizens, the physical and legal entities, directed on systematic obtaining the income, profit on using property, sales of the goods, works, rendering of services [11].

Business activity is one of types of work, is moral equal in rights with other types of work.

The church blesses any work directed to the benefit of people; thus it is not given preferences any of human activities if that corresponds to Christian ethical standards. In parables the Lord our Jesus Christ constantly mentions different professions, without allocating any of them. He speaks about work of the sower (Mk. 4. 3–9), servants and housekeeper (Lx. 12. 42–48), merchant and fishermen (Mf. 13. 45–48), the manager and workers in a vineyard (Mf. 20. 1–16) [4].

In this situation important recognition of equality of functionally different types of economic activity of people is formulated. In the given transfer are called: the businessman (merchant), the manager (housekeeper), hired workers (servants and workers in a vineyard) and free workers (the sower, fishermen) [8].

The citizens of the Russian Federation not limited in the capacity, foreign citizens, persons without nationality, and also the Russian and foreign legal entities can be subjects of business activity in the Russian Federation.

In the Russian Federation regulation of business activity is based on norms of civil law unlike the majority of the foreign states, where business activity is regulated by norms trading (commercial, economic) is right.

The right to occupation by business activity is fixed in article 34 of the Constitution of the Russian Federation according to which everyone has the right to free use of the abilities and property for enterprise and other economic activity not forbidden by the law [1, Art. 34, item 1].

The relations between the persons, who are carrying out business activity, are regulated by the civil legislation.

The civil code of the Russian Federation defined the rights of citizens to be engaged in business activity without education of the legal entity [2, by Art. 23], and also the right to create the legal entity is independent or together with other persons. The legal entities being the commercial organizations can be created, in particular, in the form of economic associations and societies, production cooperatives, the state and municipal unitary enterprises [2, Art. 48, Art. 50].

In Russia business activity is often identified with the concept «business». Though both these concepts are closely connected, at them nevertheless different sense.

The term «business activity» is meant by occupation, trade, commerce. The businessman is the man, aspiring to make profitable the activity.

The word «business» has some values. It not only business, but also any business, purchase, commercial or manufacturing enterprise, policy of the certain businessman or whole firm [5, p. 8].

Business — concept wider, than business activity as commission of any individual single commercial transactions belongs to business, in any field of activity, directed on obtaining the income (profit).

In the legislation the word «business» isn't used, but the term «business activity» is widely used [15].

Small-scale business — the business which is carried out in small forms, leaning on business activity of business owners, small firms, small enterprises. Small business is characteristic for some types and modes of production, trade, a services sector [11].

Subjects of small-scale business are understood as the commercial organizations in which authorized capital share of participation of the Russian Federation, subjects of the Russian Federation, the public and religious organizations (associations), charity and other foundations doesn't exceed 25 percent, the share belonging to one or several legal entities, not being subjects of small business, doesn't exceed 25 percent and in which average number of workers for the reporting period doesn't exceed the following limits (small enterprises):

1. in the industry — 100 people;
2. in construction — 100 people;
3. on transport — 100 people;
4. in agriculture — 60 people;
5. n the scientific and technical sphere — 60 people;
6. in wholesale trade — 50 people;
7. in retail trade and consumer services of the population — 30 people;
8. in other branches and at implementation of other kinds of activity — 50 people.

Subjects of small business are understood also as the individuals who are engaged in business activity without education of the legal entity [in 3, Art. 3].

Innovations — [fr. Innovation; lat. mnovatio — updating, change] — an novelty, qualitatively new invention [10].

In the dictionary «Scientific and technical progress» the word «innovation» means result of the creative activity directed on development, creation and distribution of new types of products, technologies, introduction of new organizational forms [7].

Innovations — novelty in the field of equipment, technologies, the work and management organizations based on use of achievements of a science and the best practices, and also use of these novelty in the most different areas and spheres [11].

Known domestic scientist R.Fatkhutdinov considers expedient to differentiate the concept «novelty» and «innovation».

Novelty — the issued result of basic researches, applied researches, development or experimental works in any field of activity on increase of its efficiency.

Novelty can be made out in the form of opening; inventions; patents; trademarks; improvement suggestions.

Innovation — the end result of introduction of an novelty for the purpose of change of object of management and receiving economic, social, ecological, scientific and technical or other type of effect [13, p. 15–16].

It is difficult to disagree with words of R.Fatkhutdinov: «Results of the last years testify to emergence of some tendencies of improvement of an economic situation in the country. However, analyzing the 15-year period of market transformations to the Russian Federation, it is with regret compelled to note that the majority production and socio-economic indexes worsened. Having huge natural resources, considerable research-and-production and personnel potential, Russia tails after the world community on quality of life, labor productivity and efficiency of use of resources, quality of the goods, competitiveness of various objects and to other indicators» [13, p. 7].

The most important factor of increase of efficiency of use of capacity of the country now is ensuring quality (competitiveness) of administrative decisions which, as we know, are developed by economists and managers. Therefore ensuring competitiveness of heads, economists and managers is the main condition of increase of competitiveness of the organization. Especially it is important in connection with accession of Russia in the World Trade Organization which mechanism of functioning sharply increases the global world competition [13, p. 8].

In August, 2009 during an informal meeting of the president of Russia with leaders of the Duma fractions, D.Medvedev urged to change economy structure: «Russia needs advance, this movement while isn't present. We mark time, and it was shown accurately by crisis». Other exit how to leave from a raw orientation of economy, according to Medvedev, at Russia isn't present. «So it is impossible to develop further. It is the deadlock. And crisis put us in such conditions when we should make the decision on change of structure of economy. Otherwise our economy has no future», — quotes Medvedev ITAR-TASS [16, 17].

In 2008, in Yekaterinburg, the group of scientists of the Ural Institute of Business of a name of Ivan Aleksandrovich Ilyin (A.Vetoshkin, N.Karateeva, A.Minyaylo) presented the system experience of religious and philosophical comprehension of special type of socioeconomic way first in domestic literature and a way of the production which main goal is spiritual increase and improvement of the person: spiritual and moral economy [6].

Now practically to all it is clear that Russia is in the deepest crisis which is characterized by system character and coverage width. Its prime cause is crisis spiritual. As a result of social cataclysms of the XX century, a separation from roots and the traditional bases of life our people lost consciousness, In the majority it has vague idea of influence of spiritual laws on economic and social practice. Society lost understanding of vital value of spirituality in orthodox tradition that negatively affects social and economic, political and legal and sociocultural development of Russia.

The approach to understanding of crisis and ways of its overcoming can be different. From the ordinary party crisis is represented, in its external manifestations, as falling of a level of production in economy, deterioration in all spheres of society: culture and education, statehood and sense of justice, political life and social security. But in the deep bases crisis can be considered as a turning point, as the recovery moment.

In the economic plan it means process of transfer settling the resource such as society managing, with the developed industrial way of functioning economy, — to arising society with new way of life, higher degree [6, p. 23].

What is the spiritual and moral economy? Let's emphasize especially that it is the highest type of managing of the person spiritualized by Orthodoxy, societies living under spiritual laws. Let's allocate main issues for the characteristic of economic development of new type.

First. Economy including spiritual and moral, isn't predominating in crisis overcoming, but only means in spiritual increase and rescue of the person. And all conversations that in the beginning it is necessary to solve economic problems and then to solve spiritual, are harmful. And if to be expressed strictly, to some extent and harmful. Economy, in effect — only one aspect, one measurement of human activity. To understand economy truly, it is necessary to understand its relativity. The economy as a certain way of managing is not the purpose, and a resort of the person, as spiritual and moral being, on the basis of social practice, in the course of his reproduction and activity. It is under defining influence of the spiritual factors, more important which were incomparably — religions, cultures, ethics, moral and so on.

Secondly. Social and economic processes of the end of the XX beginning of the XXI century in the countries of the world community showed that without the solution of the main task of radical change such as managing, the social and effective economy in the prevailing majority of the countries of the world won't manage to be created. In final documents of the economic conferences which have been carried out under the aegis of the United Nations to Rio de Janeiro (1992), Johannesburg (2002) it was noted that all of an increasing gap between poor and rich, ensuring social security needy it is possible to solve overcoming only on ways of creation of spiritual and moral economy.

Thirdly. For transition to spiritual and moral economy it is necessary to mobilize all available resources functioning in existing economy — natural, financial, human, intellectual, information and spiritual. These resources are in hands of representatives of the existing industrial economy which thinking was naturally formed in its bosom, with a certain character for this type of managing and way of life. The potential of this way is settled. Transition to the spiritual and moral economy which sprouts only start to sprout, will need the new people spiritualized by Orthodoxy. Only they can create a future economy as spiritual and moral economy which essentially differs from the existing. Already now it is necessary to cultivate such people. Also it is a task not only Orthodox Church, but also the state, and first of all educations.

In the fourth. There is a need of radical revision of all base economic categories inherent in nowadays existing market economy. It is necessary for Russian people to look from orthodox positions at the organization of new type of an economy — spiritual and moral in the root basis [6, p. 3–5].

If to talk about transition to new type of managing, to spiritual and moral economy, it is necessary to allocate two interconnected streams. The first stream is a stream of spiritual making, aspiration to a reconstruction of Sacred Russia. We should follow a way of spiritual updating, spiritualize our life, to be capable to create the second stream — creation of new type of an economy.

These two streams are interconnected. The first is predominating, it forms a basis and promotes emergence of a spiritual and moral economy. But also new type of managing as necessary means and a condition, promotes spiritual increase of the person.

In these purposes it is necessary to carry out spiritual and moral break to strong Russia.

This society is characterized by prevalence of spiritual and moral values of life over the utilitarian and material; aspiration to the highest degree of moral and erudition, careful attitude to the person as persons. In such society all resources are used for the benefit of the person consisting in his spiritual development, realization of its creative potential. The personality here is characterized by the developed feeling of the Homeland, high patriotism, readiness of protection of the Fatherland, sacrificial love to near, in particular to old men and children [6, p. 9–12].

Society of a spiritual civilization also is characterized by the developed science, health system and sports, absence in the state of system of game business and all phenomena contradicting spiritual revival and increase of moral level of the nation. It possesses the original symphony of the power and Church, the zealous relation to natural resources, their savings for the subsequent generations, unlike the market economy cultivating escalating material requirements and cultivating consumer society that conducts to blasting bases of material welfare of the nation, production senseless, and often and harmful products of production.

Society of a spiritual and moral civilization at the heart of the economic activity is guided by spiritual and moral economy. The main kernel of such society is the family. A problem of all and first of all, the states — its daily strengthening, material support, especially large families [6, p. 12–13].

Foremost problem of social sciences, scientific to give a theoretical basis and justification of spiritual and moral society, and conceiving intellectuals — to lead the people of Russia to new tops of spiritual and moral development. The spiritual and moral climate in our Fatherland to the best will change, we it is natural, we will leave on a trajectory of development of spiritual and moral economy. Theoretical-economic and practical thought should develop in this direction. This is the main way of development of Russia.

Changing spiritual and moral climate in our Fatherland to the best, we it is natural, we will leave on a trajectory of development of spiritual and moral economy [6, 12–13].

The VII World Russian National Cathedral (December 2002) created and opened «The arch of moral principles and rules in managing». They are based on ten precepts, God-given, on historical experience of their assimilation by Christianity and other religions which are traditionally professed in Russia. These principles and rules shouldn't be perceived as literal reproduction of the bible text. They should become spiritual and moral installation and a social ethical standard, starting with God's precepts in their wide understanding and concrete refraction in socioeconomic experience of our people. It is the peculiar ethical manifesto revealing an ideal model of managing and the socio-labor relations.

Without being the normative document, it, nevertheless, can become a public moral and ethical reference point of managing and work.

Moral principles and rules of managing are reduced to the following:

1. Without forgetting about a daily bread, it is necessary to remember spiritual meaning of the life. Without forgetting about the personal benefit, it is necessary to care of the benefit near, the society and fatherland benefit.
2. Wealth — not end in itself. It should serve creation of worthy human life and the people.
3. The business ethics, fidelity to this word helps to become better both to the person, and economy.
4. The person — not «constantly working mechanism». It needs time, for rest, spiritual life, creative development.
5. The state, society, business, should care together of worthy life of workers, and furthermore about those who can't earn the living. Managing is a social and responsible kind of activity.
6. Work shouldn't kill or cripple the person.
7. The political power and the power economic should be divided. Participation of business in policy, its impact on public opinion can be only transparent and open. In economy there is no place to corrupt officials and other criminals.
8. Appropriating another's property, neglecting property the general, without rendering the worker for work, deceiving the partner, the person breaks the moral law, harms to society and.
9. In competitive fight it is impossible to use lie and insults, to maintain defect and instincts.

It is necessary to respect property institute, the right to own and dispose of property. It is immoral to envy wellbeing near, to encroach on its property [6, 12–13].

The world entered the third millennium. The XX century advanced mankind by the way of mastering by new scientific knowledge and technologies. The world penetrated by telecommunication communications as though became it is less, barriers between the countries and people fall.

The market economy should be based on a principle of an equivalent exchange by the goods and services. Process of such exchange is regulated by the relations between supply and demand. The modern market economy in any country is difficult and multidisciplinary that demands state regulation. The state can't be away from market processes, but it can't and shouldn't substitute a market mechanism completely. The main objective of economic development of a market economy just also consists in creation of the balanced system within which the state only would correct negative market factors and tendencies, directed the general development to the course of satisfaction of interests of society, didn't neglect managing development [6].

The market is the most strict arbitrator and the most democratic regulator of economic selection of productions useful to society as a result of which action the average level of efficiency of national economy as a whole steadily raises.

Concrete historical conditions of development of our society testify to preservation in its economic sphere of high level of monopolism.

Monopoly (from Greek μονο — one and πωλέω — I sell) — an exclusive right to implementation of any kind of activity (production, trade, application, use of certain objects, products) provided to only a certain person, a group of persons, the state [11].

Alternative to monopolism is the open economy, development of mutually advantageous forms and civilized methods of production competitiveness among producers of the goods and services.

Not competition but cooperation and healthy competitiveness should become an essential condition for building regulated, socially oriented and civilized market. It acts as a universal form of the development of a healthy market economy [6, page 283].

The competition (from lat. concurrentia — collision, rivalry) — competition between producers (sellers) of the goods, and generally — between any economic, market subjects; fight for sales markets of the goods for the purpose of obtaining higher income, profits, other benefits [11].

The innovative approach to conducting modern business assumes introduction in processes of production and service of more and more perfect technologies and carefully developed quality systems and the service, consumers answering to constantly growing inquiries.

Application of innovations serves the small commercial enterprise as the admission on the new branch and regional markets.

At the same time with it is necessary to consider that the modern economy constantly demands correlation of coordination of results of work of each separate firm, each certain businessman and branch of a national economy as a whole.

It is possible to get quickly enough profit, having applied innovative technologies, but thus sharply to worsen a macroeconomic situation (for example to cut down the wood; to dump poisonous production wastes in a reservoir; to reduce number of workers, having replaced them with machines). But the main content of managing consists not in maximizing profit, and in care of the person, the worker and the consumer, in honest satisfaction of his legitimate and reasonable needs.

It is right to say that the world community is currently in dire need of creating a more optimal model of market relations, in the framework of which oligopoly harmoniously interwoven would with the interests of small and medium enterprises and the requirements of free competition. Such organic syntheses hardly think in Russia in the nearest future. However, de-monopolization of production and creation of conditions for healthy competitiveness on the basis of cooperation and mutual help are a main direction of economic development [6, стр.285]. It is hard to overestimate the ability of the entrepreneur to raise the competitiveness of your enterprise at the expense of a more flexible approach to the customers, improving the quality of the manufactured product and the number of additional services, effective management of staff and use of modern and moral practices.

The spiritual and moral aspect of conducting business activity means that application of innovations in work of the small commercial enterprise, for the purpose of increase of its efficiency and competitiveness, — should be directed on creation, and shouldn't lead to destructive consequences.

R.Fatkhutdinov gives three definitions connected with competitiveness:

Competitiveness — ability of the subject to be the leader, successfully to compete to (compete) to the competitors in the concrete market during concrete time on achievement of the same purpose.

Competitiveness of firm — ability of firm to let out competitive production, advantage of firm in relation to other firms of this branch within the country and beyond its limits.

Competitiveness of the manager — advantage of the manager in relation to other manager, being characterized ability to develop system of ensuring competitiveness of this object to operate collective on achievement of the purposes of system [14, p. 413].

The modern economic dictionary supplements [11]:

Competitiveness of the goods — ability of the goods to meet the requirements of the competitive market, to inquiries of buyers in comparison with other similar goods presented in the market.

At the present stage of technological revolution of firm aspire to increase specific weight of the innovations realized in innovations that allows them to raise monopolism level in this sphere and to dictate to buyers and competitors the policy. Welfare of society is defined not by mass of factors of production and not volume of investments, and efficiency of the innovative activity yielding the end result [by 13, p. 17].

Thus everything is appreciated ability of managers to identify the contents and value of service products above, to define their distinctive characteristics and to choose such system of their production, granting and consumption which would allow to get competitive advantages steady in the long-term plan.

Let's sum up.

Now the importance of a valuable approach more and more increases in Russia to scientific planning of business activity and introduction of innovations.

The valuable approach assumes social responsibility, orientation to the future, correlation of the purposes of business activity with principles of Christian morals.

The religion is included into the basis of human history, its culture and an economy, creating the most significant and highest ideals for society. The religious belief is immanent to the person communication with real Truth, with the unconditional beginning and the center of all existing, with absolute life. It is impossible to understand spirit and culture of the people, its socioeconomic system of life if not to know his religious belief and ideals.

From a position of orthodox dogma the main ethical standards with reference to an economy and social organization are considered in concepts of good, freedom, worthy existence, happiness and sense of human life [6].

The main condition for improving the efficiency and competitiveness of the organization is to ensure the effectiveness of its leaders, economists and managers.

In aspiration to increase the efficiency of their own business through innovation, head of small commercial enterprise should be guided on spiritually-moral aspect of conducting business activity, which can be formulated in several points:

The first. Orientation to people. People are the most vital resource of the organization. High compensation of the personnel, development of wide social programs. Investments into environmental protection, care about physical and moral health of the employees;

The second. Aspiration to honest profit: spiritual and moral principles of business mean the offer of qualitative products and services to the clients, at economic reasonable prices;

The third. Refusal of aspiration to receiving profit at any cost, unacceptability of receiving additional profit in a damage to the partners, orientation to human values and spiritual and moral economy;

The fourth. Social responsibility, charity;

The fifth. Complete perception of realities of surrounding reality, increase of the spiritual and moral consciousness and deep civic stand;

The sixth. Spirit of rivalry, that is ability to achieve success in conditions of an intense competition;

The seventh. The external prospect — ability to enter the unions and to achieve support from the outside, including from key financial and state figures;

The eighth. Openness of economic activity of the enterprise to financial and tax control;

The ninth. Flexibility and ability to make decisions in the conditions of uncertainty;

The tenth. Orientation to the future;

Such is spiritual and moral aspect of conducting business activity and application of innovations in work of the small commercial enterprise.

References

1. The Constitution of the Russian Federation;
2. Civil code of the Russian Federation (part one);
3. The federal law from 14.06.1995 No. 88-FZ «About the state support of small business in the Russian Federation, article 3.
4. Bible. Books of the Scriptus of the Old and New Testament. — M: Russian bible society, 1999. — 1371 pages.
5. Aleynikov A. Business activity: Educational and practical grant / Aleynikov A. — M: new edition, 2003. — 304 pages.
6. Vetoshkin A., Karateeva N., Minyaylo A. Spiritual and moral economy. Monograph. Yekaterinburg: URGU publishing house, 2008. — 702 pages. Edition 2.
7. Gorokhov V., Halipov V. Scientific and technical progress: Dictionary / Gorokhov V., Halipov V. M: Politizdat, 1987. — 366 pages.
8. Ilyin I. About a private property / Russian philosophy of a property of the XVII-XX centuries of [Text] / Ilyin I. — SPB.: 1993.
9. Kadzhaspirova G. Pedagogical dictionary: for the student. ped. studies. institutions. / Kadzhaspirova G., Kadzhaspirov A. — M: Publishing center «Academia», 2000. — 176 pages.

10. Nikitina V. Large dictionary of foreign words. — M: LLC «House of Slavic books», 2012. — 992 pages.
11. Rayzberg B., Lozovsky L., Starodubtseva E. Modern economic dictionary. — 5-e Izd., revised and expanded — M: INFRA-M, 2006. — 495 pages.
12. Tugarinov V. values of the life and culture. — HP: Ed.-in Leningrad state University, 1960. -180pages.
13. Farkhutdinov R. Innovation management: Textbook for Universities. 6th ed. — SPB.: Peter, 2010. — 448 p.: (Series «a Textbook for Universities»).
14. Farkhutdinov R. Strategic management: textbook. — 9th prod., corr. and add. — M: Publishing house «Delo», of ANH, 2008. — 448 pages, p. 413.
15. Yakovleva A. On state regulation of small business and overcoming unemployment in Russia: Problems of modern economy / journal — N 3(31) — mode of access to the journal: http://www.m-economy.ru/
16. http://www.ladno.ru/actarch/12545.html
17. http://www.itar-tass.com/

Application of Balanced Scorecard System in the Environmental Economics

Marina Treyman
*Saint-Petersburg State Technological University of Plant Polymers,
Saint-Petersburg, Russia*

Abstract. *In the modern market conditions a rather tough competition exists, therefore companies must develop and take into consideration the strategic directions in the organization management. In the article the aspects of implementation of ecology-economic indicators of water use into the balanced scorecard systems of the enterprise and the importance of implementation of these systems.*

Keywords: *strategic management, balanced scorecard system, environmental economics, ecology-economic indicators, water use.*

In its development every enterprise should be guided by the modern strategic directions of management. As long as nowadays at any services and offers market there exists a fairly strong competition between enterprises and firms of different directions and sectors of industry, so that to achieve the aims and targets it is necessary to create a balanced scorecard system for the enterprise.

«Balanced scorecard system is an effective and universal tool of management for the consecutive directing the operations (activities, measures), groups of people (organizations, enterprises, institutes, sectors, departments, project groups) on the achievement of the common goal» [1].

According to the theory of the balanced scorecard system it can be subdivided into 4 groups [2]:

1. Financial component — these indicators show if the chosen strategy leads to the high final financial results;
2. Client component — these are indicators oriented at the satisfaction of the company's clients;
3. Business processes (internal processes) — these indicators show in what key processes the organization should achieve perfection so that to become valuable for clients. They allow detecting the processes that are deciding for the results of company's operation;
4. Component of education and development – these indicators show the influence of the human capital on the components of the company's activity.

Within the balanced scorecard system the ecology-economic indicators of water use belong to the group of indicators concerning the business-processes. This information can be represented by the following scheme:

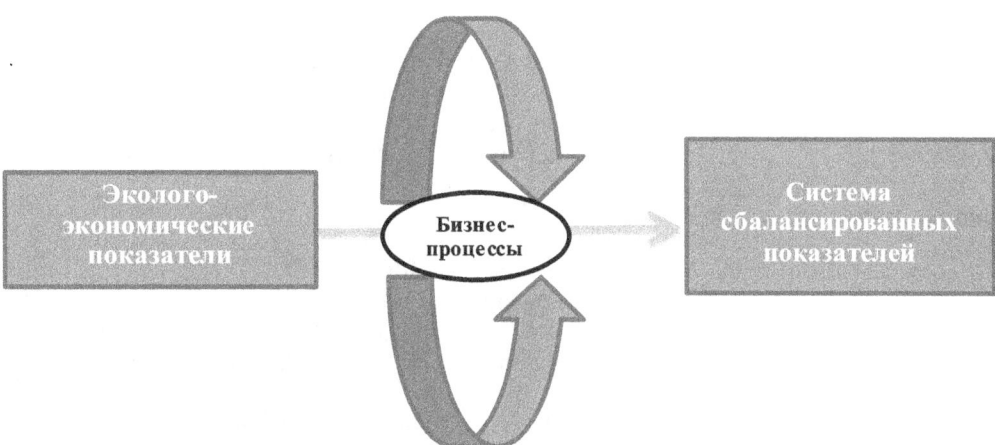

FIG. 1. Formation of the system of ecology-economic indicators of water use with application of the balanced scorecard system

Эколого-экономические показатели – economy-ecologic indicators
Бизнес-процессы – business-processes
Система сбалансированных показателей – balanced scorecard system

According to the scheme let us create a table with new ecology-economic indicators of water use with application of the balanced scorecard system.

TABLE 1. System of new ecology-economic indicators of water use

Indicator	Group in the system of balanced indicators of Kaplan — Norton	Equation or economic meaning	Possibility of usage (level, local, regional, federal)
Ecology-economic indicators of water use (water supply)			
Change of the turnover of water supply services with regards to the current period	Business-processes	Shows the effectiveness of usage of resources. Shows the share of the reduction or rational usage of water resources during this year and turnover reduction compared with the previous periods. $$\Delta Q = \frac{\Delta Q(n-2) - \Delta Q(n-1)}{Qn}$$	Локальный
Change of the turnover of water supply services with regards to the current period in monetary terms	Business-processes	Shows the reduction of income/ expenses of the enterprise (operating costs for water). ΔQ мат. исп. услуг = $$\frac{\Delta Q(n-2) - \Delta Q(n-1)}{Qn} \cdot T \text{ исп. услуг}$$ ΔQ мат. прочие = $$\frac{\Delta Q(n-2) - \Delta Q(n-1)}{Qn} \cdot T \text{ прочие}$$	Local, regional, Federal

TABLE 1. *(continued)*

Reduction of income/ expenses of the enterprise at the cost of turnover reduction	Business-processes	The cumulative reduction of income of enterprise caused by turnover reduction. ΔД пр. = ΔQ мат. исп. услуг +ΔQ прочие	Local, regional, Federal

Ecology-economic indicators of water use (water disposal)

Economic damage from disposal of sewage	Business-processes	Economic damage from disposal of sewage depends on the basic standard rate, mass of polluting substances being disposed, and the ecological situation in the region with correction for every single operating enterprise is expressed by the equation below: $y = J \cdot B (\sum Hнi + Hнi \sum 1/B - 1)$; $y = H \cdot J$	Local, regional, Federal, Global

Qфакт; Qплан - Qfact; Qplan
T прочие, T исп. услуг - T others, T providers of services
ΔQ мат. исп. услуг; ΔQ прочие - ΔQ mat. providers of services; ΔQ others
mіф - mif
mіл - mil
mф – mf
mін; mіл; mісв - min; mil; miextra
Нні, Нлі - Sn, Sl
Cін, Cіл - Pn, Pl

Conventional signs:

Qфакт; Qплан — factual and planned turnover of water supply services; in cubic meters.

Q n-1; Q n — factual turnover of water supply services for n-1; n year; in cubic meters.

ΔQ – change of turnover of water supply services with regards to the current period; in cubic meters.

ΔQ (n-2) — ΔQ (n-1) — change of turnover of water supply services with regards to the previous periods; in cubic meters.

ΔQ n — change of turnover with regards to the current period; in cubic meters.

T прочие, T исп. услуг — other tariffs for cold water supply by categories and providers of municipal services, in Russian Roubles.

ΔQ мат. исп. услуг; ΔQ прочие — reduction of income/expenses of the organization caused by reduction of operational expenses and rational usage of resources.

B =Σ mіф — in case if the disposal of sewage will be within the allowed disposal norm.

B =Σ mіл — in case if the mass of disposal of sewage does not exceed the established disposal limit.

yi — specific damage from disposal of 1 standard ton of polluting substances into the reservoir, in Russian Roubles/standard ton;

σ — coefficient of ecologic situation and ecologic importance of conditions of the water body;

j — type of polluting substance in the sewage;

Aj — coefficient of relative ecology-economic danger of j-type polluting substance, in standard ton/ton;

mф — factual mass of the annual disposal of j-type polluting substance into the reservoir, in ton/year;

mін; mіл; mісв — masses within the norm, within the limit and within the extralimit disposal of polluting substances into the water bodies, in ton/year;

Нні, Нлі — basic standard payment rate within the norm and limit, in Russian Roubles;

Cін, Cіл — payment rate within the limit and the norm, in Russian Roubles;

y — economic damage from disposal of sewage, in thousands Russian Roubles;
y — economic damage from disposal of sewage according to the new methodology, in thousands Russian Roubles;
J — correction factor of influence of the enterprise within the group of enterprises, which takes into consideration the ecologic situation in the region, specifics of the enterprise and peculiarities of polluting enterprises.

The advantages of new system of ecology-economic indicators are:

- simple and accurate calculations;
- possibility to evaluate and analyse the situation on different (local, regional, global) levels of influence;
- relevance and accounting of all the major factors of influence;
- formation of system of balanced ecology-economic indicators of water use that allows the management to make fast decisions in different financing and operating circumstances.

References

1. Herwig R. Friedag, Walter Schmidt "Balanced Scorecard: Mehr als ein Kennzahlensystem", Omega-L, Moscow 2006, 132 page.
2. Paul R. Niven, Balanced Scorecard Diagnostics, Dnepropetrovsk «Balance Business Books», 2006, 247 p. (pages 30–33; 117; 126–139)

Cardinal Voting: the Way to Escape the Social Choice Impossibility

Sergei A. Vasiljev

Micron, JSC, Novosibirsk, Russia

Abstract. *In the article it is called in question the universality of the ordinal theory of social choice as a whole and Arrow's impossibility theorem in particular. It is shown that a voting exists that cannot be described on the base of ordinal theory, and to describe it the cardinal point of view is demanded. In absence of cardinal formalization of basic axioms of the social choice theory it is offered new formal mathematical machinery. It is proved that cardinal voting can satisfy Pareto efficiency, independence of irrelevant alternative, unrestricted domain, and at the same time it can be nondictatorship.*

Keywords: *Arrow's Impossibility Theorem, ordinal voting, cardinal voting, refereeing, manipulability of a voting.*

The modern theory of social choice is based mainly on so-called ordinal point of view, starting point of which is the definition of preferences as binary relations of a kind "better - worse". Other point of view known as cardinal one is based on the measurability of a preference. The cardinal approach is criticized on the ground that "there is no meaning relevant to welfare comparisons in the measurability of individual utility" [1].

The absence of the well-founded argumentation in reply to ordinal criticism did not encourage the development of the cardinal theory of social choice. Moreover, the ordinal theory having been developed in works of Arrow [1], Gibbard [2], Satterthwaite [3] and many others pretends to universality and generality of its conclusions. The basic and the most discouraging conclusion of the ordinal theory is the impossibility of the ideal voting system, in other words, democracy is not the ideal. Kenneth Arrow had proved the theorem [1] well known today as Arrow's "impossibility theorem" (formally, the "General Possibility Theorem"). "The essence of this theorem is that there is no method of aggregating individual preferences over three or more alternatives that satisfies several conditions of fairness and always produces a logical result. Arrow's precisely defined conditions of fairness and logicality have been the subject of scrutiny by other scholars. However, none have found a way of relaxing one or more of these conditions that results in a generally satisfactory voting system immune from the voting paradox. Thus Arrow's theorem has the profound implication that in many situations there is no fair and logical way of aggregating individual preferences -- there is no way to determine accurately the collective will of the people" [4]. The set of Arrow's conditions includes: Pareto efficiency, nondictatorship, independence of irrelevant alternative (demanding that social choice over any set of alternatives must depend on preferences *only* over those alternatives), and unrestricted domain (requiring that social preference must be a complete ordering, with full transitivity, and that this must work for every conceivable set of individual preferences).

Gibbard [2] and Satterthwaite [3] independently proved the theorem that, for at least three alternatives, every Pareto optimal and non-manipulable (strategy-proof) choice rule is dictatorial. The aggregation rule refers to as manipulable if a voter, whose behavior is considered to be rational, can show false preferences rather than sincere ones to secure a more favorable outcome.

Notice that Arrow's impossibility theorem is about preference rules, whereas that of Gibbard and Satterthwaite is about choice rules. A choice rule selects one alternative for every voter profile, while a preference rule assigns to each voter profile a collective order of preference of the alternatives. Nevertheless, Satterthwaite [3] showed that "a one-to-one correspondence exists between every strategy-proof voting procedure and every social welfare function satisfying Arrow's requirements".

Frequently in the literature the Condorcet's majority criterion is used to estimate the aggregation rules. According to this criterion, if the alternative A is preferable to B for the absolute majority of voters, it should be preferable to a society. However, as it was showed (e.g. see Arrow [1]), this criterion results in paradox of cycling and cannot be considered as universal one unlike Arrow's and Gibbard-Satterthwaite's theorems.

The pessimistic conclusions of the ordinal theory, apparently, stimulated investigations of the positive decisions. Except for the mentioned above attempts of relaxing one or more Arrow's conditions, the attempts are undertaken to find the decision on cardinal base.

Smith [5] and Hillinger [6] offered, accordingly, the range voting and the evaluative voting. Both used the cardinal approach. At range voting voter is allowed to expose any rating to alternative on a uniform continuous numerical scale within limits (-1, +1). The social ranking of alternatives is defined by comparison of the sums. Hillinger [6], proving that the continuous scale is impracticable, offers a discrete scale. On pragmatic grounds, he argues for the three-valued scale (-1, 0, 1) for general elections, and the five-valued scale (-2, -1, 0, +1, +2) for elections in a committee of experts.

Both Smith and Hillinger argue that their aggregation rules meet the Arrow's conditions and at the same time they don't deny manipulability of their rules.

Another cardinal voting known as approval voting (see Brams and Fishburn [7] for an overview) is the procedure in which voters may vote for (approve of) as many of the alternatives as they wish. One can see that approval voting is a particular case of the range voting or evaluative voting with double valued discrete scale (0, 1). The approval voting is known since the mid1970s. This is why the approval voting gained numerous, but not unanimous, criticism. Thus Brams and Fishburn [7] find that approval voting is among the least manipulable of voting systems while Saari [8] and Niemi [9] find that it is among the most manipulable. Tabarrok [10] proves that approval voting is inconsistent while Cranor [4] comes to the contrary conclusion. According to de Swart et al. [11] the approval voting does not satisfy the Pareto condition and again Cranor [4] shows the reverse.

It is easy to prove that these three cardinal types of voting do not satisfy the independence of irrelevant alternative from the ordinal point of view. Given three alternatives a, b, c and three voters V1 and V2 and V3, let's consider two profiles of individual preferences that differs each other by preferences relative to a and b, from the ordinal point of view. Let on the first profile for V1 a be not worse than b, and b - not worse than c, i.e. $a \geq b \geq c$, for V2 - $a \geq b \geq c$ too, and for V3 - $c \geq b \geq a$. On the second profile let's consider unchangeable preferences of V1 and V2 and preferences of V3 - $b \geq c \geq a$.

TABLE 1. Approval voting. The first profile of individual preferences.

	V1	V2	V3	Σ
a	1	1	0	2
b	1	0	0	1
c	0	0	1	1

with outcome $a > b$.

TABLE 2. Approval voting. The second profile of individual preferences.

	V1	V2	V3	Σ
a	1	1	0	2
b	1	1	1	3
c	0	0	1	1

with outcome $a < b$.

Note that ordinal preferences of second voter do not change. Both (1, 0, 0) and (1, 1, 0) corresponds to $a \geq b \geq c$.

TABLE 3. Range voting and evaluative voting. The first profile of individual preferences.

	V1	V2	V3	Σ
a	1	1	-1	1
b	0	0	0	0
c	-1	-1	1	-1

with outcome $a > b$.

TABLE 4. Range voting and evaluative voting. The second profile of individual preferences.

	V1	V2	V3	Σ
a	1	1	-1	1
b	0	1	1	2
c	-1	-1	0	-2

with outcome $a < b$.

Note again that ordinal preferences of the second voter do not change. Both (1, 0, -1) and (1, 1, -1) corresponds to $a \geq b \geq c$.

One can see that in both cases to prove the dependence of irrelevant alternative we have to change numerical values of preferences of voters, i.e. we do not take into account the degree of preferences.

Without this changing the proof does not work. Apparently, here is a core divergence of the ordinal and cardinal point of view. As we remember, from the ordinal point of view there is no sense in the measurability of preferences. Besides, it is insisted that social choice theory does not require the assumption of cardinal utilities and ordinal description is sufficient to establish the fundamental theorems of welfare economics [12]. To prove the reverse one should give c onvincing arguments. Here is given such an argument. It is argued that ordinal theory is not sufficient one. It is proved that there do exist and, by the way, are used in practice for a long time voting systems that can not be described within the ordinal theory of the social choice and to describe it the cardinal point of view is demanded.

The matter concerns refereeing of sports competitions like gymnastics, figure skating, diving and so on. In the absence of objective criteria to rank sportsmen the collective decision of referees is used. I remind that a referee can value a sportsman numerically within some scale limits. Then referee's marks are summarized and sportsmen are ranked in accord with gained sums.

Let's consider the analog of a refereeing with scale limits (0, 1). Given three alternatives a, b, c and three referees, the Table 5 presents six possible profiles of referee's marks (preferences).

TABLE 5.

	The outcome is $a > b > c$					The outcome is $a > c > b$			
	R1	R2	R3	Σ		R1	R2	R3	Σ
a	1	0.6	0.1	1.7	a	1	0.6	0.1	1.7
b	0.5	0.1	1	1.6	b	0.5	0	1	1.5
c	0	1	0.5	1.5	c	0.1	1	0.5	1.6
	The outcome is $c > a > b$					The outcome is $c > b > a$			
	R1	R2	R3	Σ		R1	R2	R3	Σ
a	1	0.5	0.1	1.6	a	1	0.5	0	1.5
b	0.5	0	1	1.5	b	0.5	0.1	1	1.6
c	0.1	1	0.6	1.7	c	0.1	1	0.6	1.7
	The outcome is $b > a > c$					The outcome is $b > c > a$			
	R1	R2	R3	Σ		R1	R2	R3	Σ
a	1	0.6	0	1.6	a	1	0.5	0	1.5
b	0.6	0.1	1	1.7	b	0.6	0.1	1	1.7
c	0	1	0.5	1.5	c	0	1	0.6	1.6

The outcomes of given orderings differ each other. However, using the ordinal description all these orderings look equally (see Table 6), i.e. for the referee 1 *a* better than *b* and *b* better than *c*, for the referee 2 *c* better than *a* and *a* better than *b*, for the referee 3 *b* better than *c* and *c* better than *a*.

TABLE 6.

R1	R2	R 3
a	c	b
b	a	c
c	b	a

Obviously, it is impossible to find a simple function that would give six different outcomes for a single profile[1]. So any ordinal description of a refereeing violates the condition of unrestricted domain. Only cardinal description does not violate this condition.

This example shows that ordinal theory of the social choice is not sufficient to describe all possible and existing aggregation rules and the cardinal one is required to describe at least few of them. But fundamental axioms and theorems of the social choice use the ordinal formalization. So to apply the cardinal approach to the refereeing new formalization of basic conditions mentioned above is necessary.

Let there be a set of alternatives *A* and a set of voters *N*. To take into account a degree of preference let's define a cardinal preference as follows. If a referee *i* of *N* prefers alternative *a* to alternative *b* (*a, b* of *A*) it is written as $f_i(a,b) = c_i(a) - c_i(b)$, where $c_i(a)$ and $c_i(b)$ are the marks of a referee *i* to *a* and *b*, respectively. Thus a value of preferences can be $f_i(a,b) \geq 0$. The case $f_i(a,b) > 0$ corresponds to the ordinal strict preference "*a* better than *b*", and $f_i(a,b) = 0$ corresponds to the ordinal indifference.

Since $f_i(a,b)$ always has distinct value that either "equal to zero" or "more than zero" the notion of a weak preference is unnecessary at the cardinal approach.

Individual cardinal preferences are *transitive* if for any *a, b, c* of *A* and for any *i* of *N* from $f_i(a,b)$ and $f_i(b,c) \Rightarrow f_i(a,c)$.

Individual cardinal preferences are *complete* if for all *a, b* of *A* and for all *i* of *N* $f_i(a,b)$ or $f_i(b,a)$ exists.

A set of all complete and transitive preferences of all voters is called a *profile* of preferences.

Let an aggregation rule be a function *F* that assigns to each profile *p* a collective preference relation on *A*. Let's denote as *F(a,b)* the weak social preference *a* to *b*, i.e. *a* not worse than *b*, and as $F^S(a,b)$ the strict social preference *a* to *b*, i.e. *a* better than *b*[2].

A function *F* is said to be **Pareto efficient** if for any pair *a, b* of *A* for all *i* of *N* from $f_i(a,b) > 0 \Rightarrow F^S(a,b)$.

A function *F* is said to be **independent of irrelevant alternative (IIA)** if for any pair *a, b* of *A* for all *i* of *N* and for any pair of profiles *p* and *p** $f_i(a,b) = f_i^*(a,b)$ then $F(a,b) \Rightarrow F^*(a,b)$, or $f_i(b,a) = f_i^*(b,a)$ then $F(b,a) \Rightarrow F^*(b,a)$.

A function *F* is said to be **dictatorial** if there exists an individual *i* such that, for all *a* and *b* of *A*, $f_i(a,b) \Rightarrow F^S(a,b)$ regardless of preferences of all individuals other than *i*.

One can see that taking into account the measurability of alternatives tells essentially upon the independence of irrelevant alternative mainly. An ordinal preference has singular dimension - the direction ("better" or "worse"), while a cardinal preference is defined both by the direction and by the value. So while the Arrow's condition IIA demands invariability of the direction of the relevant preference, the cardinal IIA demands the preference value to be invariable as well.

Let's consider now the voting that we use in Table 1. If all individual preferences are defined they are complete. Being real numbers they are transitive. The social rank of alternative is defined by the sum of individual ranks. If individual preferences are complete then the sum exists and social

---

[1] The number of possible referee's profiles giving different outcomes for the same ordinal ordering can be much more depending on scale fragmentation, and number of voters and alternatives.

[2] In contrast to individual preferences we do not demand the social preferences to be cardinal ones. Although at refereeing the resulting sums of referee's marks are used to rank sportsmen, the ranking itself can be an ordinal ordering.

preferences are complete too. Also being real numbers they are transitive.

Pareto efficiency. $f_i(a,b) > 0 \Rightarrow c_i(a) = c_i(b) + \Delta_i$, where $\Delta_i > 0$ for all i. Then $\Sigma c_i(a) = \Sigma c_i(b) + \Sigma \Delta_i > \Sigma c_i(b) \Leftrightarrow F^S(a,b)$.

Independence of irrelevant alternative. Let's consider profiles p and p^* such that $f_i(a,b) = f_i^*(a,b)$ for all i. Then by definition $c_i(a) - c_i(b) = c_i^*(a) - c_i^*(b)$. Since the social preference is defined by subtraction of the sums of individual ranks $\Sigma c_i(a) - \Sigma c_i(b) = \Sigma(c_i(a) - c_i(b))$, then $\Sigma(c_i(a) - c_i(b)) = \Sigma(c_i^*(a) - c_i^*(b)) = \Sigma c_i^*(a) - \Sigma c_i^*(b)$. Or $\Sigma c_i(a) - \Sigma c_i(b) = \Sigma c_i^*(a) - \Sigma c_i^*(b)$ and $F(a,b) \Leftrightarrow F^*(a,b)$.

Nondictatorship. Let a be socially preferable to b, i.e. $F^S(a,b)$. Let n be a dictator. Then n can change social preferences to $F^{S*}(b,a)$ by changing his (or her) preferences relative to a and b. The highest possible such a change can be if $c_n^*(a)=0$ and $c_n^*(b)=1$. By definition, individual preferences of other voters are arbitrary. So assume that $c_i^*(b)=0$ for all i other than n, and at least one voter sets $c_i^*(a)=1$. Then $\Sigma c_i^*(a) \geq 1$ and $\Sigma c_i^*(b) = 1 \Rightarrow F^*(a,b)$, that contradicts to $F^{S*}(b,a)$ and n is not a dictator. Since a and b and n is taken arbitrarily, the reasoning is right for all available alternatives and for any voter, and so the voting is not dictatorial.

Thus a cardinal voting like refereeing satisfies unrestricted domain, and Pareto efficiency, and IIA, however it is not dictatorial in contradiction to the conclusion of the Arrow's impossibility theorem. Sen [13] affirms "that even weaker forms of (interpersonal) comparability (of utilities) would still permit making consistent social welfare judgments, satisfying all of Arrow's requirements". Hereby it is given else one confirmation of this thesis.

According to Satterthwaite [3] the manipulability of a voting is a consequence of violation of the Arrow's conditions. Since the cardinal voting satisfies all required conditions it must be strategy proof, but obviously it can allow manipulations.

Vasiljev [14] proves that manipulability of a voting depends on the available information on the outcome of a voting rather than on violation of the Arrow's conditions. The proof is done on the ordinal base. Nevertheless let's show that the proof is valid for a cardinal voting.

Any cardinal voting can be considered as an ordinal choice rule by means of substitution of alternatives. Let there be a set of alternatives A. In accord with a voting procedure there exists a set of available permutations of individual preferences on A. When voting, a voter chooses the only and the most preferable combination from this set. So, available combinations of individual preferences also can be considered as alternatives. For example let's consider a two-alternative evaluative voting with three-valued scale [-1, 0, +1]. In this case a voter is allowed to set any of these values to alternative a and to alternative b. At that the number of combinations is equal to nine: three combinations of indifference $a = b = -1$, $a = b = 0$, $a = b = +1$; three combinations of strict preferences $a > b$, i.e. $a = +1$ and $b = 0$, $a = +1$ and $b = -1$, $a = 0$ and $b = -1$; three combinations of strict preferences $b > a$, i.e. $b = +1$ and $a = 0$, $b = +1$ and $a = -1$, $b = 0$ and $a = -1$. When choosing one of them a voter shows in fact his ordinal preference. In relation to these alternatives both Lemma 1 and Theorem 2 [14] are valid.

From this point of view a cardinal voting differs from an ordinal one only by a number of alternatives. In a cardinal voting the more discretization of a scale of individual values the more alternatives are available to vote for when this voting is converted to ordinal choice rule. And for continuous scale like in the range voting this number reaches infinity.

In summary, as it is known to disprove a theory the only fact is enough that contradicts the theory. Any refereeing using numerical values disproves the assumption of the ordinal theory that "there is no meaning in the measurability of individual utility". This calls in question both ordinal criticism of the cardinal theory and the universality of one of the main conclusions of the ordinal theory – the social choice impossibility. A cardinal voting can satisfy all Arrow's conditions and can be strategy-proof at the same time and as a result it can overcome the social choice impossibility.

References

1. Arrow, Kenneth J. Social Choice and Individual Values, 2nd Ed. New York: Wiley, 1963.
2. Gibbard, Allan F. Manipulation of Voting Schemes: A General Result. Econometrica, July 1973, 41(4), pp. 587-601.

3. Satterthwaite, Mark A. Strategy-Proofness and Arrow's Conditions: Existence and Correspondence Theorem for Voting Procedures and Social Welfare Functions. Journal of Economic Theory, April 1975, 10(2), pp. 187-217.
4. Cranor, Lorrie F. Declared-Strategy Voting: An Instrument for Group Decision-Making. PhD thesis. December, 1996 Saint Louis, Missouri. The article can be download at http://users.erols.com/aejohns/node4.htm
5. Smith, Warren D. Range Voting. December 2004. The article can be download at http://math.temple.edu/~wds/homepage/rangevote.pdf
6. Hillinger, Claude. Voting and the Cardinal Aggregation of Judgments. Discussion paper 2004-09, Department of Economics, University of Munich. The article can be download at http://epub.ub.uni-muenchen.de
7. Brams, Steven J. and Fishburn, Peter C. Approval Voting. Boston: Birkhauser, 1983.
8. Saari, D. G. Sucseptibility to Manipulation. Public Choice. 1990, 64, pp. 21-41.
9. Niemi, R. The Problem of Strategic Behavior under Approval Voting. 1984, American Political Science Review, 78, pp. 952-958.
10. Tabarrok, Alexander. President Perot or fundamentals of voting theory illustrated with the 1992 election. Public Choice. 2001, 106, pp. 275-297.
11. Swart, Harrie de, et al. Categoric and Ordinal Voting: An Overview. The paper is an extended version of the original Dutch booklet 'Verkiezingen', published in 2000 by Epsilon Uitgaven, Utrecht, The Netherlands.
12. Albert, Michael and Hahnel, Robin. A Quiet Revolution in Welfare Economics. Princeton University Press, 1991.
13. Sen, Amartya. The Possibility of Social Choice. American Economic Theory. 1999, 89, N3, pp. 349-378.
14. Vasiljev, Sergei. (unpublished manuscript April 1, 2008) 'Manipulability of a Voting', SSRN 1118627, available at http://ssrn.com/abstract=1118627.

Construction Waste Management in the Republic of Bulgaria (Legislative and Practical Aspects)

Gena Tsvetkova Velkovska
Trakia University, Stara Zagora, Bulgaria

Introduction

Environmental protection is one of the main directions of the state policy of the Republic of Bulgaria.

Environmental protection is an extremely important mission for the protection and health of the people, by providing a sustainable living environment for present and future generations of the Republic of Bulgaria. Therefore, one of the main pillars in the environment is rational waste management including construction waste.

Within the meaning of Art. 57 of the Law on Environmental Protection, waste management is carried out in order to prevent, reduce or limit the harmful effects of waste on human health and the environment is ensured through:

- prevention or reduction of waste production and the extent of their risk through:
  a) the development and application of technology, providing the rational use of natural resources;
  b) the technical development and marketing of products that are designed so that their production, use and disposal do not have or have the smallest possible share to increase the amount or the danger of waste and pollution risks to them;
  c) the development of appropriate techniques for the final disposal of dangerous substances contained in waste destined for recovery, recycling or processing;
- waste recovery through recycling, reuse or reclamation or any other process of extracting secondary raw materials or the use of waste as a source of energy;
- safe storage of waste unfit for recycling at this stage of development.

Persons whose activity is related to the formation and / or treatment of waste are obliged to ensure the recycling and disposal in a manner which does not endanger human health and use methods and modern technologies:

- not result in damage or risk to the environment components;
- not cause additional strain on the environment related to noise, vibration and odors.

Bulgaria to the European Union comply with several regulations relating to waste management, their treatment, their processing, recycling, etc. such as:

- European Union Directive on waste classification.
- Directive of the European Union — when waste ceases to be waste.
- European regulations for the transportation of waste.
- Etc.

Specific mechanisms for waste management in the Republic of Bulgaria Act are reflected in the Waste Management Act, the ratification of the Basel Convention on the control of transboundary transfer of hazardous wastes and their treatment etc. Moreover, in Bulgaria operate a number of regulations, securing legislative mechanisms for waste management in their species diversity, treatment worked, etc. For the purposes of this study below will look at some key points related to the waste management and the factual situation in the Republic of Bulgaria at this stage.

1. Management of construction waste and use of recycled building materials (within the meaning of Regulation on management of construction waste and use of recycled materials)

The text set out in this order are directed to manage different types of construction waste generated by construction works, regardless of the category of construction. In addition, this group includes construction waste and removal of constructions including after introduction into service, as well as recycled materials.

Bulgarian legislation requires before starting construction works and / or removal of construction contracting to prepare a management plan for construction waste.

With specific text, the legislature prohibits illegal dumping, incineration, and any other form of unauthorized treatment of construction waste, including disposal in containers for collection of household waste and packaging waste.

What is a management plan for construction waste in Bulgaria?

The management plan for construction waste include:

- general data for the investment project;
- description of the subject of removal;
- forecast of construction waste generated and the level of material recovery;
- forecast the type and quantity of products utilized for construction waste, which are used in construction;
- measures to be taken in the management of construction waste generated in accordance with the requirements of regulations and laws.

In the negotiation process for the award of construction works and / or removal of construction contracting authority or authorized officer:

- determined responsible for the implementation of the management plan for construction waste for the building;
- imposes obligations to the participants in construction and investment process to meet the requirements of the targets for recovery and recycling of construction waste and use of recycled materials and / or recycling of construction waste backfilling.

When carrying out construction works and removal of buildings, construction waste must be separated by type and transmitted for further material recovery.

An important requirement of the Bulgarian legislator construction waste to be collected, stored, transported and prepared for use separately.

This training is done on specially equipped sites—these are sites for recovery and recycling of construction waste.

Sites are of three types. Each of the sites must meet specific requirements. Requirements must meet construction waste:

- construction waste must comply with the requirements laid down in the investment project construction;
- person making the material recovery through the use of construction waste backfilling must have a document according to the requirements of the waste management for activities waste treatment code R10.

In which cases can be used ie construction waste when they can be recovered material? This is possible in the presence and compliance of the following conditions:

- inert construction waste in accordance with the established requirements of the Law on waste management.
- construction waste has undergone preparation prior to recovery and / or preparation for reuse.

Construction waste, suspected that do not meet the criteria for inertia and / or originate from sites that did not meet the legal requirements or other contaminated sites undergo mandatory testing as concerned laws and regulations for the management of waste. The results of the tests to inertia documented in test reports issued by accredited laboratories.

2. Control of waste management

Control of waste management, the Bulgarian legislator assigned to state and local authorities, namely:

- The mayors of municipalities.
- Director of the Regional Inspectorates of Environment and Water.
- Minister of Environment and Water of the Republic of Bulgaria
- Etc.

What constitutes those control functions of these bodies?

The mayor of the municipality or a person authorized by him shall control:

- activities of generation, collection, separation, storage, transportation, treatment of household and construction waste;
- and the disposal of industrial and hazardous waste in municipal and / or regional landfills;
- compliance with legal requirements for waste management, reflected in specific regulations, ordinances, etc.

Moreover, the mayor of the municipality organize and control, reclamation and subsequent monitoring of landfills for municipal and construction waste on the territory of the respective municipality.

Director of the Regional Inspectorate of Environment and Water (Bulgaria is divided into 28 administrative districts ie the number of regional visits, 28) or an authorized officer shall exercise control over compliance with the requirements for waste treatment and the conditions of the permit, respectively a registration document for:

- activities of generation, collection, separation, storage, transportation, waste the territory of the Regional Inspectorate of Environment and Water;
- facilities and installations for storage and treatment of waste.

Based on the violations in its inspection of the Regional Director or inspection authorized by him shall give obligatory prescriptions a deadline for their removal and / or draw acts of administrative violations.

Director of the Regional Inspectorate or an authorized officer exercising control over:

- proper calculation and timely payment of product fees specified in the relevant legal texts
- Implementation of the obligations of owners of landfills on the financing of disposal by landfill.

Minister of Environment and Water of the Republic of Bulgaria or a person authorized by him shall exercise control over:

- compliance with the terms of licenses issued to recovery organizations and persons performing individual management obligations of widespread waste;
- activities on waste management;
- compliance with EU regulation № 333/2011.

Minister of Environment and Water, along with several representatives of the central government (Minister of the Interior, the Minister of Transport and Communications and the Director of the Agency «Customs») control the transboundary movement of waste under this Act and regulations of the European Community № 1013/2006 in accordance with its powers.

This assessment is based on:

- Director of the Regional Inspectorate on whose territory the place of origin of the waste, or authorized by the officials;
- Director of the Regional Inspectorate on whose territory the destination of the shipment, or authorized by the officials;
- customs authorities, the General Directorate „ Border Police» and the units „ Traffic Police „ in the regional directorates of the Ministry of Interior;
- The officials of the Executive Agency „ Automobile Administration „ Executive Agency «Railway Administration» Executive Agency «Maritime Administration», the authorities of the General Directorate «Border Police» and the units „ Traffic Police „ in the regional directorates of the Ministry of Interior.

To the supervisory authorities, the Bulgarian legislator is associated representatives of the Ministry of Health and — right on regional health inspections. Director of the Regional Health Inspectorate and the Director of the Regional Inspectorate of Environment and Waters or authorized by them shall exercise control activities hazardous waste treatment in health facilities.

Checks required by law, can be of two types:

- inspections of documents
- spot checks are also recorded.

The Law on Waste Management provides at least one inspection per year under consideration bodies.

On-site inspection is independent and takes place once a year in place of the activity in the presence of verifying or individuals who work for him. In the absence of such persons check is conducted with the participation of at least one witness.

Inspections concerning collection and transport operations shall cover the origin, nature, quantity and destination of the waste collected and transported.

What is the scope of a check?

Officer carrying out on-site inspection is entitled to:

- access the premises in which a controlled activity;
- require the presentation of documents in accordance with the statutory requirements must be located at the place of inspection;
- require written and oral explanations from anyone who works for the inspected person;
- involve experts in the field, where the inspection is complicated and requires special knowledge.

If on-site inspection is established lack of documentation demonstrating compliance with the statutory requirements of the inspected person is determined 7 days of their submission.

When carrying out supervisory bodies up protocols and / or acts of administrative violations. When violations supervisory bodies provide mandatory instruction and set a time to eliminate the violations.

When carrying out control bodies shall protocols. When violations supervisory bodies provide mandatory instruction in ascertainment a deadline for their removal and / or acts for the establishment of administrative violations.

Control functions in waste management are also, as already mentioned, the customs authorities of the Republic of Bulgaria.

The customs authorities shall carry out customs supervision and control of transfrontier shipment of waste in accordance with the established rules of the Law on Waste Management.

The bodies of the General Directorate «Border Police» and the bodies of the units „ Traffic Police „ in the regional directorates of the Ministry of Interior shall exercise control over the transboundary movement under the requirements of the Waste Management Act, the Ministry of Interior and regulations for its implementation and can take the appropriate legal action.

Officials of the Executive Agency „ Automobile Administration „ Executive Agency «Railway Administration» Executive Agency «Maritime Administration» exercise control over the transboundary movement of waste under the requirements of the law, relevant international legal instruments ratified by the Republic of Bulgaria Act, road Transport Act, the road Traffic Act, railway Act, the maritime Spaces, Inland Waterways and Ports of the Republic of Bulgaria, the Merchant Shipping Code and regulations for their implementation and can take appropriate legal action.

Control authorities and persons exercising control, have the right of access to documents, records and other materials, as reflected in the Law on Waste Management. When violations, supervisory authorities shall draw up acts of administrative violations.

3. Some practical aspects of waste management in the Republic of Bulgaria (national report on the state and the environment in the Republic of Bulgaria in 2011 — edition 2013)

Bulgaria continues to pursue an active policy on waste management. Presented in outline this policy is reflected in those national report as follows:

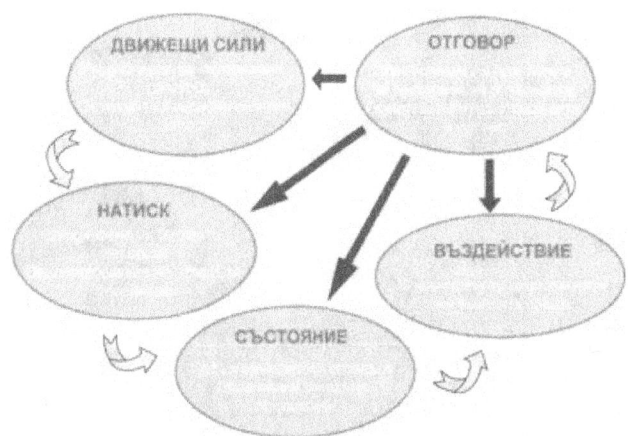

A / Hierarchy of waste management

In 2011, the quantity delivered for recycling waste 16 % of the total generated, 5 % more than in the previous 2010

In 2011, 84 % of the waste delivered for disposal including disposal.

Definition of the indicator

The indicator is the ratio (in %) of the amount of waste recycled compared to the amount of total generated per calendar year and presents the hierarchy of waste management and how to treat waste at the national level. Indicator type — response.

Rate Indicator

Disposal including Landfill is still the most — the application activity waste treatment. However, a trend of increasing quantities of waste delivered for recovery including recycling.

Source: National Statistical Institute of Bulgaria

FIG. 1. Activities waste treatment, %

In the hierarchy of activities in the field of waste as a priority indicating the prevention of waste, and break the causal link between waste generation and economic growth and environmental impact. The scheme represents the arrangement of the ways to treat waste at national level reporting / assessment of their impact on the environment.

FIG. 2.

B / Generated waste

Five-year period (2007–2011) amount of waste is reduced by 20 %, mainly due to reduced growth in construction activities in the country, mainly due to economic reasons.

Between 2007 and 2011 the amount of hazardous waste decreased by 73 %, the main decrease was in 2011. compared to 2010 in 2011. hazardous waste is 69 % less compared to the formed in 2010. The reason is the procedure fayalitov reclassification of waste from the business of the business „ Basic metals and fabricated metal products, except machinery and equipment».

Definition of the indicator

The indicator presents the amount of waste by type nationally. Indicator — type pressure.

Rate Indicator

In the Republic of Bulgaria in 2011, the total amount of waste is 15 897 kt, (202 kt hazardous and non-hazardous kt 15 695 incl 2753 kt bit). Quantity of waste (hazardous and non-hazardous including household) by type for the period 2004–2011, are presented in Fig. 3.

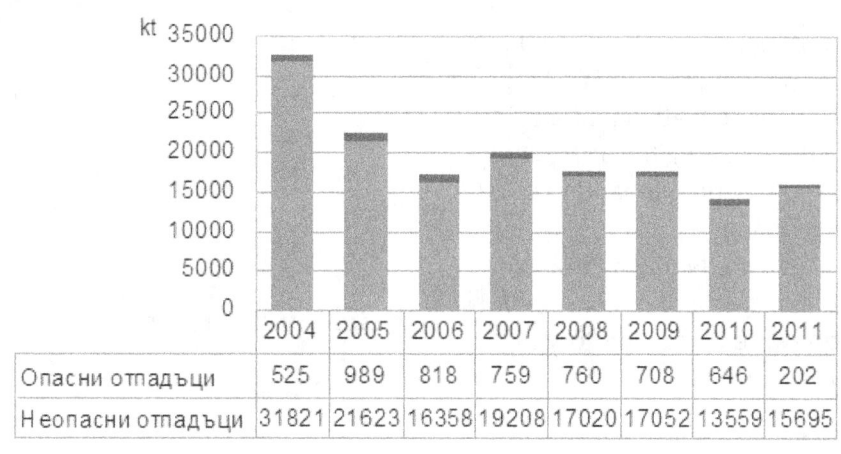

FIG. 3. Of waste by type, kt
Source: National Statistical Institute of Bulgaria and the Executive Environmental Agency of the Republic of Bulgaria

Amount of waste generated can be viewed as an indicator of how effectively society, particularly in terms of resource use and selection of the most appropriate methods of waste treatment.

One of the objectives of the European Union «break „ the relationship between the amount of waste and consumption of materials and resources, respectively economic growth.

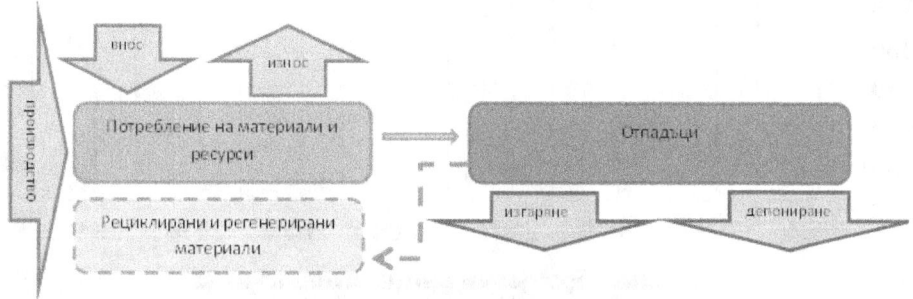

FIG. 4. Link — consumption waste
Source: European Environment Agency

Increasing consumption and developing economies continue to form large quantities of waste — as more efforts are needed to reduce and prevent their formation. Waste is an environmental, social and economic problem.

C/ O generated and treated waste

Over the past four years (2008, 2011). Amount of generated waste in the country is decreasing.

The average rate of accumulation of waste for the EU -27 for 2011 is 520 kg / year / per capita, while Bulgaria, it is 380 kg / year / per capita.

Reducing the waste generated mainly due to: Introduced administrative tools:

- Measures set out in the National Programme for Waste Management 2009 to 2013.
- Measures set out in the National Strategic Plan for gradually reducing the amount of biodegradable waste going to landfill (2010–2020). (Strategic Plan);
- Measures set out in the National Strategic Plan for waste management of construction and demolition waste on the territory of Bulgaria for the period 2011–2020

Introduce economic instruments:

- Deductions for each ton of disposed waste and increased cost of disposal;
- increase with each year of product fee for polymer bags;

Introduced practical tools:

- Providing financing for construction of regional systems for waste management;
- Provide funding for regional facilities for pre- treatment of municipal waste;
- Provide funding for the closure of municipal landfills;
- Increase households covered by home composting systems;
- Reduce population.
- Raising public awareness about the benefits of separate collection of waste and to their recovery as a valuable economic resource is also an important aspect in the management of waste and — particularly in the field of household and construction waste.

For example, in 2011 recycled 160 kt waste, compared in 2010. quantities are 26 kt.

In 2011, 2568 kt disposed waste and remains the most used method in the country for treatment nabitovi waste.

Definition of the indicator

The indicator represents the quantity of waste generated and treatment activities of household waste at the national level. Indicator — type pressure.

Rate Indicator

Municipal waste for 2011 2753 kt.

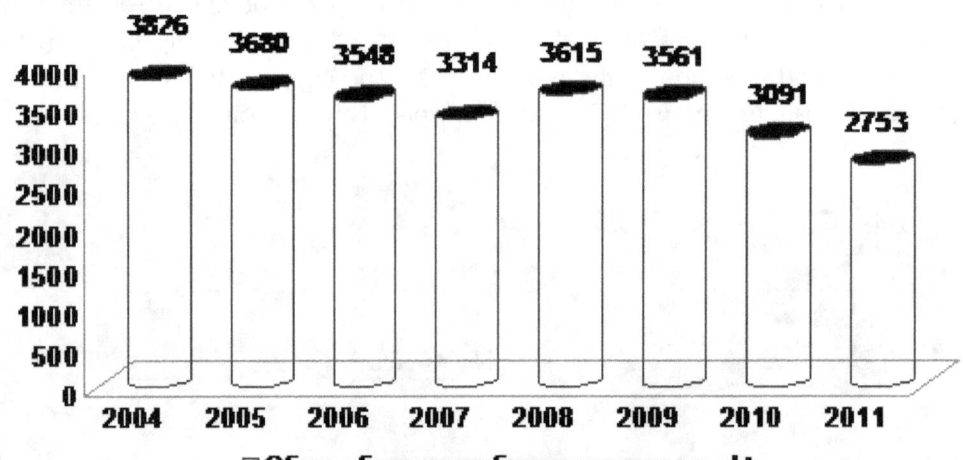

FIG. 5. Waste production, kt
Source: National Statistical Institute of Bulgaria

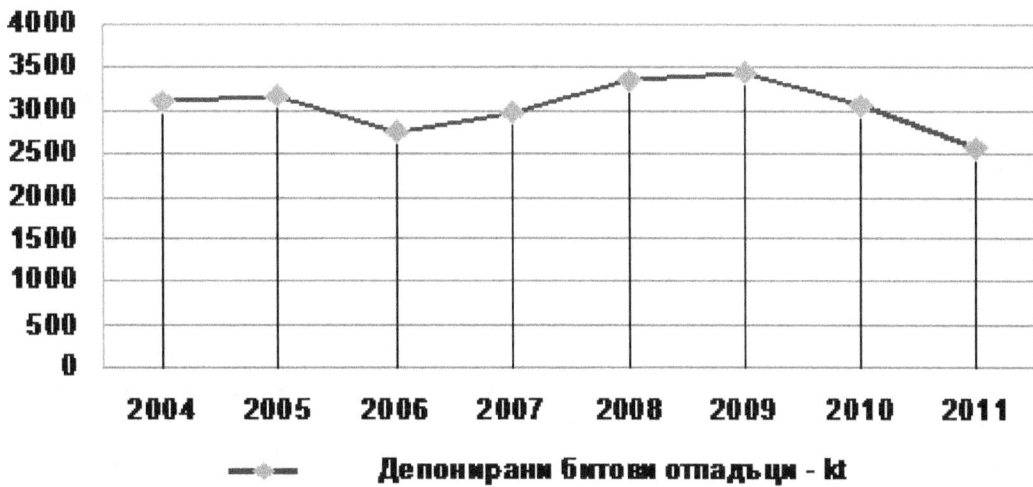

FIG. 6. Landfilled waste, kt
Source: National Statistical Institute of Bulgaria

We want to remember that waste is the waste that are produced as a result of the vital activity of people at home and in administrative, social and public buildings. Household waste have many socio-economic aspects and their complex character indicates the status of the management of these wastes at the national level. Household waste is collected in the municipalities, are the main source households.

For reference, the period 2004–2011, the observed mean of about 9 % per year in the total amount of waste generated and by about 7 % per year in the amount of landfilled waste.

Landfill as a disposal method is ranked last in the accepted hierarchy of waste management, but this method still has the highest share of municipal waste treatment. In municipal waste containing recyclable and biodegradable materials and their use is essential, and favorably affect the environment.

With support from the European Structural Funds programs and projects at the end of 2011 g.sa built 28 regional landfills for municipal waste disposal.

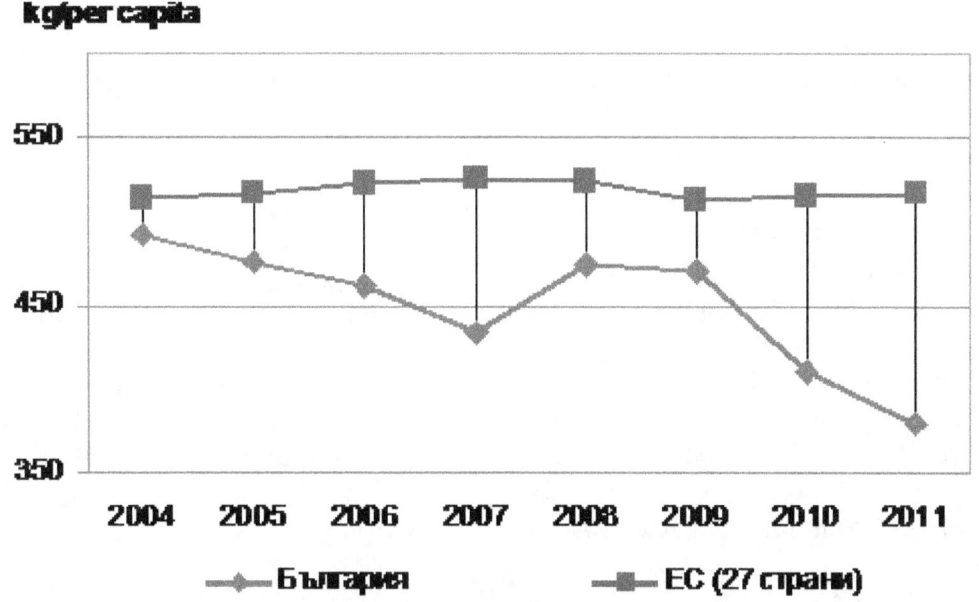

FIG. 7. Accumulation rate, kg / per capita
Source: Eurostat

The accumulation rate is the amount of municipal waste generated per year per capita. Annual amount of generated per capita waste varies as some countries have achieved stabilization of the volume of waste, or even decline, while others follow a steady increase.

References to legal and strategic documents.

Directive 2008/ 98 of the European Community on waste and repealing certain Directives, revoke the Waste Framework Directive 2006/12 of the European Community, effective 12.12.2010g.. The main objective set out in Directive 2008 /98 of the European Community is to reduce the use of resources and to promote the application of the hierarchy of waste management., So as to reduce the negative impact on the environment and to increase the efficiency of utilization of resources.

Conclusion

The long-term objective of the European Union is to become a recycling society that will prevent waste and promote the practice of using waste as a resource. European Union seeks to „ improve the efficiency in the use of resources to reduce the overall use of non-renewable natural resources and the relevant impacts of the use of materials on the environment through the use of natural resources at a rate that does not exceed their regeneration potential» (Upgrade Strategy EU sustainable Development, 2006). Preventing waste mainly done by limiting unnecessary consumption through design and consumption of products that generate less waste., prevention also includes actions that can be taken after the product reaches the end of its life cycle by recycling and recovery of waste materiali.Efektivnostta of resource use is several times higher in the EU- 15 than in the new Member States, the European Union and the countries of Southeast Europe. Projections for 2020 indicate that the use of resources in the European Union will continue to increase (Waste and material resources., The European Environment Agency „)

Under the new Waste Framework Directive to assess the impact on the environment it is necessary to apply the approach takes into account the entire life cycle of materials, which examines the environmental impacts from raw material extraction to waste generation.

Requirements of the framework legislation on waste management at European level have been introduced in the Law on waste management regulations for its implementation.

In the National Programme for Control of Waste formulated objectives and measures for waste management, which should ensure the implementation of an integrated and effective system of waste management at all levels.

The strategic plan must ensure compliance with the requirements of Directive 1999/31 of the European Union of 16 July 1999 on the gradual reduction of the amount of biodegradable municipal waste going to landfills, as authorized for disposal by 2013. — 50 % of the generated in 1995g.kolichestva biodegradable waste.

In some strategic documents of the Republic of Bulgaria lays the foundation for an integrated framework for managing construction waste to reduce the environmental impacts caused by their generation, improving the efficiency of resource use, increased responsibilities, and stimulate investments in waste management. According to the European Environment Agency, the largest waste streams in Europe come from construction and demolition.

References:

1. Law on Environmental Protection
2. Law on Waste Management
3. http://www.eea.government.bg/
4. http://eea.government.bg/bg/soer/2011
5. http://www.lex.bg
6. http://www.moew.government.bg
7. http://www3.moew.government.bg/files/file/POS/Strategic_documents/NSPUOSR-final.pdf
8. http://www3.moew.government.bg/files/file/Waste/cdw/NAREDBA_CDW.pdf

LAW

International and National Experience of Application of Ecological Audit

Victoria Erofeeva

Vasilisa Kraeva

Volgo-Vyatskiy institute (branch) of the Kutafin Moscow State Law University (KMSLU), Kirov, Russia

Abstract. *In this article the international experience of ecological auditing is reviewed. ISO standards are analysed.*

Keywords: *ecological audit, ecology, ISO standards.*

International experience shows that countries widely use ecological audit procedures as the means to get and evaluate the ecological information about the enterprise in order to develop the necessary corrective measures and make decisions on different management levels.

The US «Pollution Prosecution Act», adopted in 1990, stipulates that offenses of ecological law which have been voluntarily disclosed or amended due to the ecological auditing program do not compose a set of elements of the administrative or criminal offense. This Act authorizes the Criminal Prosecution Department with serious powers. Offenses eliminated voluntarily or amended due to the ecological auditing program do not become the subject of the criminal proceedings. Justice department has declared that all voluntary efforts for prevention of ecological offenses and for assistance to investigation are considered as the mitigating circumstances, the ecological auditing program being one of them [1]. Enhancing the ecological auditing procedures, professional training of auditors, exchanging experience and information in this field is the activity of different associations in the USA, and the largest and most well-known are «Roundtable on ecological audit», «Ecological Auditing Forum», «Formal Association of Ecological Auditors and Managers — Institute of Ecological Auditing» [2].

In 1995 the European Union has adopted the set of documents on ecological management and ecological audit (ISO 14000 standards). These standards are the first series of international standards on environment management which were developed by the International Standardization Organization. This organization was established in 1947 for assistance in elaboration of standards which would promote the development of international relationships, trade and manufactures. At present the members of ISO are national committees from more than 118 countries, including the Russian Federation [3].

Nowadays the series of the most widely-used ISO 14000 standards of the International Standardization Organization includes more than 30 separate documents and is constantly being added by new documents translated into Russian language and issued in the Russian Federation in the form of GOSTs. Conventionally, ISO 14000 standards are referred to five different groups, differentiating the sets of documents concerning the organizational and management aspects, technological processes and features of products. Both international and national ecological standards are voluntary and cannot substitute the legal provisions. However, their application allows estimating the impact made by a company on the natural environment and if the company fulfils the legal requirements according to the ecological norms, and also detecting the internal factors of the non-effectiveness of company's activities and decreasing the production costs.

Compliance to the ISO 14000 standards (as well as International quality standards ISO 9000) becomes obligatory for business in industrially developed countries however in the Russian Federation so far only a few companies are determined to pass international ecological certification. Mostly these are enterprises of nuclear and oil-and-gas industry, where the level of requirements is very high. Judging by distribution of international standards ISO 9000 it can be assumed that companies try first of all to get the formal confirmation of their «ecologically responsible conduct» and this intention will be the driving force of implementation of ecological management systems corresponding to ISO 14000 standards. The experience of several countries in the certification according to ISO-14000 shows wide possibilities of usage of environment management. For example, in Switzerland the finance company Credit Suisse Group was the first big financial institution which got certified by ISO-14000. In 1995 this company entered into the alliance with four other banks and a big insurance company and developed the «Management principles of energy consumption». These principles establish the targets in the field of environment protection, mainly what regards the energy consumption that since a long time has been acknowledged as the factor most significantly influencing the bank operations.

Since April 1997 the banks of Swiss companies have been preparing for certification, and in April 2000 the certificate has been confirmed, so the environment management system is now implemented by this company on all banking operations performed by its international branches. Credit Suisse Group is in active and open dialogue with its employees and partners; at the same time the certification allowed Credit Suisse Group to include into contracts the requirements of implementing the environment management system [4].

According to the «Closed Substance Cycle Waste Management Act», adopted in Germany in 1997, companies bear direct responsibility even in the case if the harm to environment has been done in the result of regulated exploitation of production equipment, that is, they are responsible not for the harm, but for the risk of the harm. Besides, according to the creditor's obligation the creditor can become the defendant for the environmental damage caused by his debtor [5].

The significant role in conducting of internal ecological audit belongs to government authorities. Thus, *the Federal Ministry for the Environment of Germany (Umweltbundesamt — UBA)* issues special *guides* to conduct ecological audits and thus the evaluation of ecological influence of the enterprise on the environment becomes more rational and objective.

Contents and procedures of ecological audit are not stipulated by the federal law of the Russian Federation. Nevertheless, many countries have already reaped the benefits of the relevance of ecological auditing, the research of its organization and performance in these countries can help in a certain way to fill in the gaps in legislation of the Russian Federation. Since many years in the Russian Federation the development of the draft legislation «On ecological audit» is in process. In 2003 the Ministry of Natural Resources of the Russian Federation has prepared the concept of the draft legislation «On ecological audit», and after its adoption the ecological audit will be given its own normative legal base in the Russian Federation.

References

1. Demina L. A. Ecological audit: the past, the present, the future // Energy: economics, technics, ecology. -1999.-№ 11.-P.33.
2. Makarov S. Foreign experience of development of ecological auditing activities// Auditor. -1997.-№ 10.-P.36
3. Please, see.: http://www.ecoline.ru/
4. Information bulletin. Adapted from ISO. Moscow, Russian Research Institute for Certification (RRIC), 2000, № 1, p. 2–32
5. Ibatullin U. G. Market of ecological services: ecological audit// Environmental economics. Survey information.-2001.-№ 1.-P.67

Commercial Arbitration as a Civil Society Institution

Tatyana Letuta

Orenburg State University, Orenburg, Russia

Abstract. *In the work we have attempted to understand the legal essence of arbitration as the Civil Society Institution with regard to the modern legal system development in Russia. In view of Fiduciary Theory of Judging of statistical data, produced by the foreign researchers, we evaluate the role of the specified institution, analyze the draft law on the arbitration court, generalize the judicial practice, as well as the lawyers and philosophers of Law opinions regarding the advisability of law reformation herein.*

Keywords: *arbitration, Civil Society Institutions, reformation of the judicial system, civil legislation.*

The legal reforms [9] taking place in Russia provide us an evidence of a qualitatively new stage of the State Legal System development; give rise an objective necessity of theoretical understanding of rationality for the proposed approaches in terms of Law Theory and Philosophy. All projects are aimed first of all at improving the quality of legal services at improving the quality of justice and, in general, at creating an effective system of protection of human rights and interests. However, according to the Lawyers they are trying to provide the declared objectives with improper means [7], and the numerous negative reviews of the Legal Community regarding the optimality of specific legal changes [13] cast doubt on the productive promotion of the Russian state on its path of democratization and the Civil Society Institutions formation. In this respect, today, the special relevance has the development of theoretical and legal concept of Arbitration as the Civil Society Institution [11]. This will allow us to understand the legal nature of this phenomenon and to reveal its role in the context of modern transformations.

The civil society, as a union of free and equal citizens in the form of voluntary associations, corporations and the like, which have an autonomy with respect to the State Authorities, is the foundation of Democracy. It provides the settlements of human rights and interests based on the self-government and freedom. The main advantage of the civil society is the presence of strong sources of authority existing outside the State. The democratic state is characterized by the developed civil society and the authoritative self-regulatory Civil Institutions. In such conditions the main purpose of the State is to use the power authorization and the enforcement tools for the protection of inalienable rights of citizens, and to make the civil society integrity and self-sufficiency. The latter is possible through the formation of a liberal legal regulation of civil society institutions and granting of freedom for the citizens to exercise their rights and interests. Thus, in the presence of a developed civil society development the specificity of its correlation in the mandatory with the State is revealed: the State provides the coordination of private and common interests, the mediation between the personality and the State [12]. In its turn, the civil society is concerned about maintaining the balance between the interests of the society as a whole and the interests of individual Civil Society Institutions, in particular.

The researchers have noted the Civil Society Institutions governmentalization tendency in Russia and, as a consequence, the civil society absence in the sense requested the by history, i.e. the really working, civilized system of peculiar structures [6]. There are only some separate, low-power organizations, unable to influence the life of the country. It seems that according to Theoretical typology of relations between the state, business and civil society proposed by V.L Tambovtsev, for Russian is more typical the version, in which the State dominates over the interaction between the civil society and businesses. [14] While the optimal version is in which the State, business and civil society interact with each other constructively. These theses are eloquently confirmed by the provisions contained in the draft Law "On Arbitration courts and arbitration (arbitration proceedings) in the Russian Federation" (hereinafter - the draft). When analyzing the provisions in the draft a strong belief in the excess of the state control over the activities of the institute proposed by the developers is formed, that Institution by its legal nature is the brainchild of civil society and based on the principle of voluntary treatment of the parties and the enforcement of a judgment. The appeal to the arbitration proceedings by the Parties in the terms of the democratic state and the developed civil society shall have the "competitive" advantages in comparison with the state civil process, because the first is carried out within the Civil Society Institution, and the second – by the State Authority. The arbitration is the quintessence of equality between the parties in the economic turnover, the optionality of the civil law regulation and the prevalence of contractual, partnership relations even in the presence of conflict. Such interpretation allows us to understand the main meaning and the mission of the specified institution, well as the boundaries of its legal regulation. The arbitration, in its essence, cannot and should not be created with the authoritative state command, the arbitration courts competition cannot and shall not be achieved through the state coercion and the false reduction of their number, the Arbitrators themselves cannot and shall not be appointed by the State Authorities. An Arbitrator is an authoritative independent individual with the impeccable reputation, to whom the entrepreneurs trust the settlement of their dispute that they could not settle on an amicable basis. Accordingly, the requirements to the Arbitrator and the procedure of his appointment shall be maximally exempt from the state regulation. To date, there are a lot of generally accepted criteria for the candidates for the post of judge of the State Court, as well for the selection of people to the Arbitrators' List. Thus, at the legislative level it is appropriate to fix only the following requirements: an individual, who has a higher legal education, has no had never had the criminal records, not recognized by the court as incapable, partially capable, has no diseases impeding the implementation of powers of the arbitrator may be appointed as the Arbitrator. At compliance with the requirements above, the person, who has reached the age of 25 years and has an experience in the legal specialty not less than 5 years, can be the Arbitrator. The establishment of other restrictive barriers to carry on the arbitration proceedings would have meant the direct state intervention in the 'private' sphere. In connection with this the proposed by the Draft requirements for the establishment of Arbitral Institutions are raising eyebrows among the non-profit entities that provide services on the arbitration proceedings administration on a permanent basis. The founders shall obtain the permission from the Ministry of Justice of the Russian Federation based on the Decision of the Interdepartmental Expert Council (Article 39 of the Draft). When establishing the Arbitration Court, the Ministry of Justice will check its Statute for compliance with the Law, and further control its ongoing activities within the Law "On Noncommercial organizations".

Contained in the draft proposals are seen frustratingly against the background of modern Law Theories. The progressive ideas of Law allow you to form a Fiduciary Jugging Theory even for the judges of State Courts that fundamentally changes the approach to understanding the role of judges in the democratic State [4]. The proclaimed independence principle of the judicial system, according to the authors of this theory - Ethan J. Leib, David L., does not allow you to understand the true purpose of a judge. The main component in determining the proper conduct of the judge shall not be concentrate on his duties as the representative of state authority, but on his responsibility before the members of democratic society, which have entrusted by the person to exercise of judicial power. In the given approach is clearly traced the need to move the attention of the legislator from the issues of state regulation to the issue of liability of the State representatives before the citizens in the developed civil society. Hence it seems more logical the fact that the status of the Arbitrator and the arbitration institutions shall not be checked by the State and shall be measured with the demand for the arbitral proceedings by this Arbitrator. The arbitration proceedings is nothing more than the service of the civil and law nature, therefore, the higher the trust of members of the society to the concrete Arbitrator and the arbitration institution, the greater the number of disputes shall be settled by this Arbitrator. It becomes clear that the introduction of licensing procedure to establish the Arbitration Courts will only be an additional incentive for the development of corruption component and will help to reduce the number of potentially effective Arbitration Courts.

Of course, the existing for today in Russia legal basis for the arbitration activities is not fully reflect its main purpose. The arbitration as well as the State Court provides the protection of civil rights violations (Art. 11 of the Civil Code of the RF, Art. 1 of the Federal Law "On International Commercial Arbitration", Art. 1 of the Federal Law "On Arbitration Courts"). The standards ensuring the fair trial of civil disputes usual for the State justice shall be applied as to the arbitration proceedings as to the alternative method of dispute settlement [3]. Despite such significant importance of the specified institution for the civil disputes settlement it is not possible uniquely to determine the arbitration status under the Russian legislation. Since, the arbitration is not a legal entity under the Russian legislation; the arbitrators themselves are not in an employment relationship with the arbitration. The arbitration can be established by any legal person, but provided that it is neither a branch or a representative office, or a structural subdivision or the authority of a legal entity that established this Court. The legislation provides the notification procedure to establish the arbitration (required notifying the competent State Court), the duty to inform on the termination of court's activities is recorded nowhere. There is no single Register of Arbitration in the Russian Federation.

Despite such a liberal approach of the legislator to the establishment and operation of this institution, the statistical data, contrary to popular belief, does not testify that the main reason for the unpopularity of the arbitration proceedings among the Business Community lies in such legislative connivance. For example, in a survey conducted in the city of Togliatti, to the question "Why don't you apply to the Arbitration Court", the majority of respondents have noted the following: - the lack of information about the court and the procedure of deal with cases (59%), the habit to apply to the State Court (22%), there is no need (6%) and only 13% said the partiality in the decision-making at the Arbitration Court [15]. According to the survey of the All-Russian Public Opinion Research Center (VCIOM) the trust level of the representatives of the Russian legal and business community to the Arbitration Institution is 6.3 points out of 10 possible [10]. With all its modesty the specified numeric value shows more positive attitude of respondents to the Commercial Arbitration, rather than the negative. Taking into account the fact that, according to VCIOM's data, consisting in the Trust Rating to the Major Political and State Institutions, the State Judicial System takes consistently only 9-10 place out of 14, Taking into account the fact that, according to VCIOM's data in the Rating of Trust to the major political and state institutions the state judicial system consistently takes only 9-10 place out of 14, it becomes obvious an ungrounded exaggeration of legal regulation problems of the Commercial Arbitration, to be more accurate, of its overly liberal nature [8].

As stated previously, the regulation of the activity of the arbitration is not perfect, however, the changes proposed in the draft even more set aside the above institution from the objectives and the essence of this event to make it as the product of a mature democratic society. Thus, in the explanatory note to the draft was noted, that it aims to establish a modern and efficient legal regulation mechanism and contains a number of original provisions which are intended to enhance the authority and the attractiveness of the Commercial Arbitration. As a result of the draft law adoption we suppose to reduce the load on the state courts and develop of the civil society institutions. We propose to achieve the specified objectives through the paradoxical means. For example, we propose to introduce the State Courts as the Assistance and Supervision Bodies in relation to the Commercial Arbitration in order to strengthen the guarantees of impartiality and independence of Arbitrators and to reduce the State Courts' load. Assistance and Supervision Bodies will intervene in the arbitration proceedings in cases related to the composition formation of the Arbitration Court, to the settlement of question of the Arbitration Court competence, etc. In such an approach we can see no compliance with the laws of logic. First, the purpose to reduce the load on the state courts seems unachievable, as they will carry out the additional, not inherent in them previously functions. Secondly, contrary to the draft developers' opinion, the impartiality and independence of the Arbitrator cannot be guaranteed by the State Court control, as well. This statement follows from the understanding of the following phenomenon. The entrepreneurs expressed their will to resolve a dispute by the Arbitrator - a person, who is not a judge in the State Court. However their will is not taken into consideration, the State does not leave the entrepreneurs a freedom to choose a form to settle their civil law disputes, actually bringing their dispute under the State Court control. The impartiality and independence of the Arbitrator shall be guaranteed by the parties to the dispute themselves and the Commercial Arbitration, since otherwise all Civil Law Contracts shall be controlled by the State for the presence therein of possible grounds for invalidity. The Arbitration Agreement, as the Civil Law Contract, assumes an agreement of parties to settlement their dispute in the Commercial Arbitration. The given main purpose of the Arbitration Agreement shall not be limited by the government intervention into the adjudication proceedings.

Specified in the draft objectives are successfully achieved outside the framework of government involvement in the foreign legal systems. Thus, for example, in accordance with the Commercial Arbitration Rules of the American Arbitration Association (AAA), in case the parties have agreed otherwise, when two or more plaintiffs, or two or more defendants, are involved in the dispute, the American Arbitration Association may appoint all Arbitrators (Section 7 of the Rules) [1]. Unless the parties otherwise agree or will not reach an agreement, according to the International Arbitration Rules of the American Arbitration Association (Article 7, 8.1, 9 of the Regulation) [2] the decision to challenge an Arbitrator shall be taken by the Administrator, or the AAA is entitled to decide whether the Arbitrator is subject to challenge (Section 13 of the Rules) on its own initiative. Thus, the procedural issues decision of the arbitral proceedings does not extend beyond the Civil Society Institution itself – the Commercial Arbitration, which allows us to speak about its independence in relation to the State.

One of the objectives of the Bill is the reducing the number of arbitration courts. This objective acknowledges the authoritative researchers opinions on the absence of partnership relations "the State - business - Civil Society" in Russia. The democratic state seeks to promote the Civil Society development, including through the establishment of numerous of Civil Society Institutions, the formation of simplified legislative base for the establishment of all kinds of non-governmental organizations, for the small business development, providing the self-government to the Civil Society Institutions.

Apparently, the legal essence of commercial arbitration could be disclosed in the legislation and in a softer approach to the activities control of this Institution. Thus, for example, the various forms of legal practices carry out their activities rather effectively. Under the existing strict requirements and strict accession procedure into the Lawyer Community, the Lawyer juridical activity itself is beyond the State control. The State does not interfere in matters of professional development and the like – all juridical and organizational activities are administered by the Federal Chamber of Lawyers – and the special education, not related to the number of Public Authorities.

In that case, if the reformation of Commercial Arbitration proceeds in the directions indicated in this draft, the Commercial Arbitrations operate in Russia - not the Civil Society Institutions, but some of their similarity. As it was correctly noted by E. Sukhanov, the arbitration proceedings extends an opportunity to dispute resolution in the sphere of civil turnover by public self-regulation, not turning into a proper judicial law protection, but representing the tendency to the democratic principles simplification in the justice system [4].

Amid the domination of integration processes in the world economy, the Russia's WTO accession, an increased confidence of entrepreneurs in the proper protection of their rights, strengthening the reputation of both the individual companies and the States represented by them generally depend not only on the regulatory regime, but also on the willingness of Russia to recognize for the arbitration courts as the civil society institutions of their significant role in strengthening the stability of civil turnover. The presence of publications in the legal literature on the need of strictly state control over the Commercial Arbitration merely shows a low degree of maturity of the social structure, which is, among other things, manifested in the fact that a State is not ready to the self-regulatory institutions rooting in the civil society.

References

1. Commercial Arbitration Rules of the American Arbitration Association (translated by Zakharova A.S.) // http://materialotzakharova.narod.ru/American_Arbitration_Association_Commercial_Arbitration_Rules_in_Russian.htm; http://www.adr.org/sp.asp?id=22440 (30.01.2014)
2. International Arbitration Regulations of the American Arbitration Association // http://www.alppp.ru/law/pravosudie/47/reglament-mezhdunarodnogo-arbitrazha-amerikanskoj-arbitrazhnoj-associacii.html (30.01.2014)
3. Decision of the West Siberian District Federal Arbitration Court dated July 24, 2013 in Case No. А45-4261/2013; Decision of the Constitutional Court of the RF dated 26.05.2011 No. 10-П // http://www.consultant.ru/document/cons_doc_LAW_114541/#p99 (30.01.2014)
4. Ethan J. Leib, David L. Ponet & Michael Serota. A Fiduciary Theory of Judging // CALIFORNIA LAW REVIEWB 2013. Vol. 101. N 3 http://www.californialawreview.org/articles/a-fiduciary-theory-of-judging
5. Kolodina, I. 30.01.2014. Judged by the conscience // http://www.rg.ru/2013/06/25/sud.html
6. Mitroshilova N.V. 31.08.2009. On the modern concept of Civil Society. Philosophical Problems.
7. New Arbitration Courts. 31.01.2014. The Bill of the Ministry of Justice was issued // http://zakon.ru/Discussions/novye_tretejskie_sudy_opublikovan_zakonoproekt_minyusta/9998
8. Estimation of the Courts activities in the RF // http://wciom.ru/courts/(30.01.2014)
9. Draft Federal Law No. 47538-6 On introducing amendments to the first, second, third and fourth parts of the Civil Code of the Russian Federation as well as to the Certain Legislative Acts of the Russian Federation. http://www.consultant.ru/document/cons_doc_LAW_146920/#p156 dated 30.01.2014; the Federal Law Draft on Amendment to the Constitution of the RF No. 352924-6 "On the Supreme Court of the Russian Federation and the Procuracy of the Russian Federation"// http://www.consultant.ru/law/doc/sud_ref/ (30.01.2014)
10. Press release No.2409. Institution of arbitration system in Russia: opinion of the legal community // http://wciom.ru/index.php?id=459&uid=114500(24.09.2013)
11. V. Putin instructed urgently to improve the Law on the Arbitration Court // http://www.rg.ru/2013/12/12/tret-sud-anons.html(30.01.2014)
12. Savinkov, L.L. 2007. Several approaches to understanding the term of "Civil Society". Bulletin of the Orenburg State University No.7.
13. Smolnikov, D. 23.12.2013. It's obviously the unconstitutional reform. The Courts Union was criticized in the Public Chamber // http://zakon.ru/Discussions/eto_yavno_nekonstitucionnaya_reforma__v_obshhestvennoj_palate_podvergli_kritike_obedinenie_sudov/8738
14. Tambovtsev, V.L. 2007. The State as the initiator of Civil Society development. Social Sciences and Modernity No. 2, pp. 69-77.
15. Chubarov, V. 30.01.2014. The third will be... // http://www.rg.ru/2012/11/13/treteiski.html

Achieving Social Consensus as a Strategic Goal of the Russian State

Aleksey Treskov

Southern Federal University, Rostov-on-Don, Russia

Abstract. *The article explores the necessity of achieving the social consensus in the Russian Federation, it being due to a high level of the social tension and proneness to conflict. Such consensus, in its turn, provides a means of achieving the social unity, alleviating the social tension and, consequently, facilitating the formation of the civil society. The necessity of achieving the social consensus is determined by the fact that the ultimate goal of any modernization lies in ensuring a sustainable growth. These transformations are primarily accomplished by the people themselves, by their public engagement and leadership ability, their political and legal involvement in the institutions of the civil society of Russia.*

Keywords: *civil society, rule-of-law state, social consensus, leadership ability, public engagement, law, theory and history of state and law, political science.*

Establishing the social consensus provides an important goal of the development of the Russian state. The contemporary period of our state is characterized by a steadily increasing social tension and proneness to conflict. Currently Russia undergoes a systemwide collapse occurring alongside with emerging conflicts which affect, actually, all fields of the civil and state functioning. The Russian state as well as the shaping civil society functioning under modernization is under strenuous conditions of the lack of consensus in basic values and interests. The necessity of achieving the social consensus is determined by the fact that the ultimate goal of any modernization lies in ensuring a sustainable growth. This goal in the multinational Russia with its confessional variety is related to achieving the social peace. November 7 may have been announced as Day of Consensus and Reconciliation with the aim of bringing the society to stability. To achieve these goals one must solve a good deal of related tasks, which will make it possible to develop the principles of legal responsibility, political unity and social consensus. These transformations are primarily accomplished by the people themselves, by their public engagement and leadership ability, their political and legal involvement in the institutions of the civil society of Russia.

The social consensus suggests a unity of viewpoints and actions to tackle major public problems. While carrying out reforms and transformations, one must remember that the formation of the civil society is related to an evolutional, non-violent development, a forward-bound movement. Undoubtedly, establishing an open civil society and ensuring the functionality of its institutions are labour-intensive and take a long time. The civil society normalizes the social life of the country by means of its institutions, creating a firm basis for that. The experience of the political and public events of the previous century has revealed the futility of violent approaches towards the social organization of public life. The only solution is to consolidate people, their abilities, who can avert the imminently increasing crisis of civilizations, to unite people in the conditions of different worldviews.

The final stage of the development of public connections means the state which essence lies in the fact that "all its citizens have and acknowledge, apart from various and private interests and goals, the integrated interest and integrated goal, namely: the common interest and common goal, because the state stands for a certain spiritual community" [1]. The state cannot exist without the common interest, universal goal or without solidarity. The solidarity is one of the major indicators of the civil society, "the true and real basis of the state" [2]. It is this basis that the state can build on as a living system of brotherhood. In other words, the state cannot exist without the civil society united by common ideas and interests [3]. I.A. Ilyin speaks about the rational meaning of the life of free and voluntary unions in a unified domineering union, in the state [4]. The coercive power of the state and diversity of its activities under correct organization not only fail to suppress a free spiritual life of a human being, but also they create favourable conditions for that. The interest of the state consists in ensuring natural rights of the citizens and their exercising such rights [5].

The fundamental element of any state power, i.e. the key governing device, which is of more significance that the army or the police, is the ideology. N.A. Berdyayev points out that the ideology lays down the system of priorities and values according to which the people live and which determine their conduct. The domination is by far based on mesmerizing words. The state resorts to indoctrination permanently. According to N.A. Berdyayev, "if people were deprived of the ability to get affected by hypnosis, it is not known, what power could maintain itself" [6]. At present the core of the problem of the Russian state lies in the fact that the main idea in the contemporary society is not clearly defined. The lack of the national idea means a spiritual crisis of the society that is characterized by a gap of the integrated space, loss of unanimity of opinion of the basic social values, decreased national self-esteem [7]. At the same time, according to N.S. Bondar, with the fundamental social values, fundamental basics and principles of the organization of the society and the state recognized at the highest level, in many respects despite the constitutional ban on the enforcement of a statutory or compulsory ideology, a natural process of rendering the relevant principles, ideas and concepts, value-laden normative standards as law-significant is bound to take place, which is a natural reveal of the constitutionalization developing in a pluralistic worldview society [8].

However, the suggested national idea of "the post-socialist Russia", namely, the point of view by V. S. Nersesyants as "the civilism as a national idea", is worth noting. In a number of his works V.S. Nersesyants theorizes about the socialism, social property and social law in terms of civilizational positions by overcoming the excesses of the world-order approach. On the whole, according to him, the civil property and civilism provide "the Russian idea today and for the future, the Russian contribution to the global historic progress of freedom and equality of people" [9]. The civil society as a society of property owners creates objective prerequisites for the formation of the social consensus. The social and political order of this society may be referred to as the civilism.

Undoubtedly, the civil society contains all the prerequisites of the social consensus. Equally, the social consensus is a mandatory condition for the formation of the civil society and a must for its successful functioning. In this respect the ideological component of the social consensus which is able to brace the society with a spiritual solidarity must be used to capacity. In other words, the ideology is not a mere reflection of a real event as the society but constitutes an essential part of this entity [10].

The social consensus is the power of the world's significance that is prepared by the previous stage of history of the human development. The social consensus must be of imperative nature which reveals in the fact that under contemporary conditions the consensus suggests resorting only to non-violent forms of the enforcement of justice. The non-violent forms are wide-scale and suggest an active engagement on the part of all individuals involved in a conflict. This was mentioned by the famous American philosopher John Rawls. In his opinion the civil disobedience is a public act. It addresses not only public principles, it is carried out publicly. It suggests an open participation with an honest notice, not a clandestine one. For this reason, among others, the civil disobedience is a non-violent act. The civil disobedience suggests a disobedience to the law within the scope of loyalty to the law, although being strained to its limit [11].

The imperative nature of the social consensus for the Russia of the XXI century is also proved by the fact that the social consensus is immanent with democracy. However, the budding democracy provides the base of the enforcement of the will of the majority which, according to J. St. Mill, is only the will of those who manage to get recognized as the majority. The dominance of the majority presupposes the ignorance the interests of the minority and sometimes the suppression of and violence over those. Such democracy is flawed and leaves much to be desired. The true democracy always seeks the universal consensus [10].

S.L. Frank in his work "The spiritual fundamentals of the society" explores the civil society as a specific subject matter. According to Frank, the social life of people is organized within two connected original principles - conformity to the plan and spontaneity, i.e. the society is created by two ways - consciously and in an orderly manner or spontaneously, inadvertently. "These two indispensable and correlative original principles of public existence get expressly revealed in two forms of public life: in the state and in the civil society" [12]. S.L. Frank suggests a personal view on the civil society as based on the functional approach. "An individualistic moment within the structure of the society, according to which the society is supposed to be divided into a number of separate, mutually independent specific centres of active forces, is not the goal of public life, but only a function, being a necessary one - a supraindividual goal of the society as a unity" [12]. The essential conditions of the civil society are freedom and cooperation among people. According to Frank, it is the interest of serving the society rather than a legal individual right that constitutes personal liberties. The personal liberties as well as the independence of people are prerequisites for the public consent and unity. "The civil society is the public solidarity emerging spontaneously of free cooperation, free consent, free wills of individual members of the society" [12].

The uprise of the personality conscious of his or her rights and aspiring after freedom and dignity is a social and economic prerequisite for achieving the social consensus. The personality shapes the industrial society where self-regulation is perceived as the freedom most generally understood to be "consisting in the fact that freedom is a definition of the man not from outside, but from inside, from the spirit" [13].

The situation in Russia as it has developed initiates the following steps:

- - changing the concepts of the society that would correspond to the transitional period from the industry-oriented to the post-industry/information-oriented stage. The perception of the society as a cultural organism where the national identity appears as the main idea becomes crucially significant;
- - restoring the independence of the moral reasoning as a form of life-oriented philosophy. A transition to the strategy of generating a "moral majority" is essential;
- - supporting the continuity of the federalization of the country, mobilizing intellectuals, particularly, first thing those pertaining to the humanities, with the aim of building up a new perception in the public mind of the cultures of the multinational nations of Russia and the integrated spiritual space.

The political stability of the ruling authority complying with the interests of the majority appears as an important stage of moulding the social consensus and fundamental strength of the society. With the way of living of people changed, their attitude towards the ruling authority and to the requirements put forward by the latter changes, too. In course of the development of the state authority it is essential to proceed from the continuously valid rules, to get guided by the verifiable requirements of the ruling majority and scale-up the forms of power in accordance with the trend of the national form of the social consciousness of the people.

The achievement of the social consensus is primarily referred to the function of the civil society; consequently, the way to the rule-of-law state lies within the formation of the communication relations between the civil society and the political one - the state.

The dynamic solidarity of the civil society and the state needs remarkable efforts on both parties for the continued existence of that. Under these conditions, along with the political, economic and social factors the role of the spiritual and moral tools which are covered first and foremost by the function of the formation and maintenance of the social and spiritual community "we are Russian nationals" is essential. The social consensus provides an expression of the deep phenomenon related to the very pivots of life of any multinational society. These pivots characterize the inner property of an integrated social organism which stands for the multinational population of our country [14].

The Constitution of the Russian Federation lays down the principles of the sovereignty of the people, priority of the human rights and liberties, rule of law, political diversity etc. Thus, in accordance with Art. 1 of the Constitution of the Russian Federation Russia is a democratic federated rule-of-law state with a republican form of government. Pursuant to Art. 2 of the Constitution every human being, his or her rights and freedoms are regarded as the supreme value. The recognition of the rights and freedoms of the human being and the citizen, their observance and protection are the obligation of the state. According to Art. 3 the holder of the sovereignty and the only source of power in the Russian Federation is its multinational people. In line with Art. 13 of the Constitution the establishment and the activities of public associations which goals are aimed at a violent change of the basics of the constitutional order and the violation of the integrity of the Russian Federation, the undermining of the state security, the formation of armed units, the incitement of the social, racial, national or religious hatred are prohibited.

Thus, it appears from the above legislative provisions that the social consensus is one of the supreme values of the state which goal suggests being in the service of the human being. At the heart of this value is the security thanks to which individuals are supposed to develop without any restraint.

The transformations taking place in the 90ies of the XX century are the reason for the disparity of the people in terms of income and quality of life. The majority of the people were disappointed at the results of the reforms in the economy and state governance in terms of economic and political freedoms, which resulted in an open conflict proclaiming individual and collective requirements as lawful. At present the civil society in Russia needs the stability both in the political and in the legal field, support of the principles of a fair distribution of the whole range of natural resources, created means of production, material benefits, valuables being at the dispose of the country as well as stimulation of the efficiency of the economy. In our opinion, in order to mitigate the acuteness of the social conflicts in the Russian Federation as much as possible one needs the social support-oriented economy with a culture component of the rule-of-law state along with the tools designed to search for public compromises.

In the western part of the world it has long been well established that the social stability prevails over the statutory or private interests. The political consensus understood as a social one is a complicated phenomenon with its own structure and contents. An excessive strive for compromises reveals in the activities of the government and legislation bodies in terms of their mutual relations. The majority principle accounts for the activities of the parliamentary democracy; for this reason the declaration of this principle often turns into an obsession for many governments and parliamentary factions. Establishing a parliamentary majority and setting up a coalition government will be the common denominator. The political course elaborated by this government turns out to be a compromise. It shall be noted that the majority principle is a mandatory principle of the democracy because it helps resolve most of the conflicts. Nowadays the problem of the social consensus in Russia is still relevant. Unfortunately, we have no stable trend in its formation. The rule-of-law state automatically assumes the strongest factor of the conciliatory processes in the society. A powerful principle of the influence of the consensus on public life is ensured by such a Constitution which meets the principles of the democratic society, preserving the inviolability of the proprietary right, stability and development of the middle class. Article 85 of the Constitution of the Russian Federation sets forth that the President of the Russian Federation may use conciliatory procedures for resolving the discrepancies between the state government authorities of the Russian Federation and administration bodies of the government authorities of the constituents of the Russian Federation as well as those between the state government authorities of the constituents of the Russian Federation. The general consensus failing to be achieved, the President may refer the dispute for settlement to court. As an example of the efficient conciliatory procedures under the auspices of the President of the Russian Federation one may set the negotiations on achieving common goals in the activities of the federal government authorities, government authorities of the constituents of the Russian Federation, parties, trade unions, public and religious organizations, which resulted in the Agreement on Social Accord opened for signature on 28 April 1994. The Agreement was signed by more than 800 signatories. It was aimed at stabilizing the political situation in the country [15]. The Edict of the President of the Russian Federation of 1 September 2000 (as in force on 10.08.2012) established the Council of the Russian Federation assisting the President of the Russian Federation, while using the conciliatory procedures [16]. Regretfully, the Agreement of 28 April 1994 was rejected by the opposition. This Agreement is regarded to have ceased to be in force with the commencement of the military hostilities in Chechnya that were undertaken without consent of the Agreement signatories and both parliament chambers.

We presume that it is necessary that more practical open public discussions, adversary positions etc. be introduced. The currently available debates cannot replace in full the true polemics in the society. The opinions addressing the government structures and contradicting the stance of the majority of the media appear in the press more and more; therefore the opposition viewpoints keep up shifting into a freer medium - the Internet.

With a view to achieving the social consensus Russia needs open conditions for discussions, those being independent television channels. Thus, Edict of the President of the Russian Federation No. 455 of 17 April 2012 "On Public Television in the Russian Federation" was officially published on 19 April 2012 to the effect that, as proceeding from the wording, the television media shall promptly, accurately and comprehensively inform the citizens of the Russian Federation of the current affairs in internal and foreign policy, culture, education, science, spiritual life as well as in other fields. However, it is supposed to be the initiative of independent social structures.

The political middle ground appears as another stage of achieving the consensus in Russia. The problem of achieving the consensus of opinion within the country, that of the course of its development, that of the trend in challenges, of the ways to cope with the latter becomes more important. The opposition supporters grow in number, more and more people participate in the meetings against the ruling government authority. It is impossible to reach the social consensus without resolving this issue. It appears that the incumbent authorities need either a drastic change in the development programme or an opportunity to be granted to other political forces elected by a direct declaration of intent to define the further development of the state. In order to achieve this mutual understanding, make the relations in the society more tolerable it is crucial that the structure of the authority institutions be reviewed. To this end the formation of the civil society which makes the power accessible and its activities transparent is essential. The ability to get organized independently, to get united for the sake of common interests is already revealing the civil society in Russia. Numerous political actions of the citizens of the Russian Federation that commenced after the elections to the State Duma (Parliament) of the VI convention, taking place on 4 December 2011 and continuing during and after the campaign dedicated to the election of the President of Russia held on 4 March 2014 are an example of that. According to the research results published by Levada Centre, 46% of the Moscovites support the protest meetings, 25% being against them. 73% approve of the demands of the protesters to punish all those guilty of falsification and 71% are in favour of investigating the evidence of violations during the elections [18].

In the opinion of L.Yu. Grudtsyn and S.M. Petrov, the modern Russian state can hardly be referred to as a strong one. It socialist predecessor was by far stronger. The weaknesses of the new Russian state are accompanied by the amorphism of the civil society and that of the process of its formation. The civil society proper is to a considerable degree replaced by its political imitation generated by the new government establishment in the form of pseudopublic organizations set up by such officials. Moreover, the political events of the 4-7 December 2011 (the elections to the State Duma (Parliament) of the sixth convention and the demonstrations of the opposition and dissenters at Triumph Square on 6, 10 and 24 December following the announcement of the results as well as the mass protest demonstrations in other regions of the country) make someone sit up and take notice of possible positive trends and a gradual replacement of the "political imitation" and pseudopublic institutions by the fully functional institutions of the civil society. The "Triumph Square" syndrome in Russia is most likely to turn into a long-lasting process, which, undoubtedly, is bound to enhance the consolidation of the civil society [19].

Head of the Council for the Development of the Civil Society and Human Rights Mikhail Fedorov declared at the final meeting that "we have managed to stir up the public consciousness, creating the conditions for the people to get transformed into the civil society by getting involved into the public partnership... we are convinced that several thousand strong meetings and demonstrations are a normal form of the civil activity, by no means the only one, though. We are convinced that the steam of the public energy must work not only for the honk, but also for the development of the country" [20].

There is a close link between the civil society and social partnership tracked down. The latter helps ease the tension within the social groups. The fundamental principles underlying the civil society require that in defiance of the discrepancies as to the available development goals the achievement of the consensus as regards the fundamental values be acknowledged as essential for the purpose of settling those differences. In this respect no means of force or compulsion shall be applied; this requires not only a high level of the legal and political culture, but that of the general one. Only then may this refer to cooperation, mutual assistance and understanding.

While studying the history of the XX century, one can see that the social consensus is difficult to reach in terms of social relations, moreover internationally.

Thus, certain goals of the social consensus in resolving global problems were achieved in the second half of the XX century. But still there remain the problems of demographic, military and terrorist nature with outlining visible positive social consensus trends: the United Nations Organization, the European Union, the European Court of Human Rights, the Organization for Security and Cooperation in Europe, which gradually results in the global consensus.

The results of the analysis of the current condition of the social consensus in the Russian state bring to the conclusion that the social consensus provides such a system of social relations between its constituents (citizens, public organizations, state) that is subordinated to the collective vitally important goals.

The importance of achieving the social consensus in the Russian Federation is determined by a high level of the social tension and proneness to conflict. This consensus, in turn, provides a means of achieving the social unity, alleviating the social tension and, consequently, it facilitates the formation of the civil society in Russia.

Reference

1. I.A. Ilyin. "The Way to Obviousness". M., 1993. P. 362.
2. I.A. Ilyin. On Essence of Legal Consciousness // Theory of state and law a / Edited & prefaced by V. A. Tomsinov. M., 2003. P. 267.
3. E.A. Parasyuk The idea of the civil society in the Russian legal science of the late XIX - early XX century // Journal of Russian Law. 2010. No. 12. P. 101– 109.
4. I.A. Ilyin. Theory of state and law / Edited & prefaced by V. A. Tomsinov. M., 2003. P. 124.
5. I.A. Ilyin. On Essence of Legal Consciousness. P. 270.
6. N.A. Berdyayev. The Realm of Spirit and the Realm of Caesar. M., 1995. P.315.
7. Speransky. The Patriotism and the National Idea // Urals Federal District. 2003. No. 5. P. 44–45.
8. N.S. Bondar. The judicial constitutionalism in Russia in the sense of the constitutional justice. M.: Norma, Infra-M 2011. P. 27.
9. V.S. Nersesyants. The Civilism as the National Idea // The Independent Newspaper. 1995. 1 November; The civilism as the concept of the post-socialism: freedom, law, property // Social Sciences and Modern Times. 1995. No. 5.

10. V.V. Semiletenko. The social consensus as a factor of stabilizing social systems: Dissertation for the degree of the candidate of philosophical sciences. M., 1999.
11. J. Rawls. The theory of justice / Academic editing and preface by V.V. Tselischeva; Translated from English. 2nd Edition. M., 2010. P. 322.
12. S.L. Frank. Spiritual fundamentals of the society. M., 1992. P. 136.
13. N.A. Berdyayev. The Fate of Russia. M., 1990. P. 283.
14. E.A. Mikulchinova. The social consensus as a factor of stabilizing the transforming Russian society: Dissertation for the degree of the candidate of social sciences. Ulan-Ude, 2004.
15. Commentary on the Constitution of the Russian Federation (clause by clause). 2nd Edition, revised and amended. / Edited by L.A. Okunkova: BEK, 1996.
16. Edict of the President of the Russian Federation of 01.09.2000 № 1602 (as in force on 12.03.2010) "On the State Council of the Russian Federation" // The Russian Newspaper, No. 172, 05.09.2000.
17. Edict of the President of the Russian Federation of 17 April 2012 No. 455 "On Public Television of the Russian Federation"// The Russian Newspaper, Federal Issue, No. 5759, 19 April 2012.
18. Yury Levada Analytical Centre (Levada-Centre) [Electronic resource] // The Moscovites about protest meetings // Access mode: http://www.levada.ru/22-12-2011/moskvichi-o-protestnykh-mitingakh, free. Screen title – Rus, Eng.
19. L.Yu. Grudtsyna, S.M. Petrov. The ruling authority and the civil society in Russia: mutual influence and contradictions // Administrative and Municipal Law. 2012. № 1. P. 19–29.
20. Dmitriy Medvedev held the final meeting of the Presidential Council for Human Rights // The Russian Newspaper. Metropolitan Issue, No. 5769 (96), 2 May 2012.

POLITICAL SCIENCE

The Partnership of the Republic of Uzbekistan with UN for the Problems of Central Asian Regional Security

Oybek Abdimuminov
Navoi State Pedagogical Institute, Navoi, Uzbekistan

Abstract. *Given work analyses the cooperation of the Republic of Uzbekistan with UNO and its agencies in the following sphere: regional problems, safety, and stable development in Central Asia. In the opinion of researchers, this cooperation speaks of Uzbekistan's international authority.*

Keywords: *Central Asia, external policy of the Republic of Uzbekistan, United Nations Organization, UNO agencies, Central Asia, regional problems, problems of Afghanistan, problems of Aral Sea, relationship, cooperation with international organizations, regional stability in Central Asia.*

Introduction

Central Asia is an important region in international affairs. The Central Asian republics became a member of the United Nations Organisation in March of 1992 at the 46th Session of the General Assembly of the United Nations. For the last eleven years, close contacts have been established with all main structures of the Organisation. The United Nations Office opened in Central Asian republics the following year. Today, there are more UN programmers, funds and agencies operating in Central Asian. The UN system in Central Asian countries works as one, collaborating to support the national reform efforts. Central Asian states cooperate intensively within the framework of the UN General Assembly agenda and with the various specialised institutions of the UN system. In the framework of the United Nations, Uzbekistan put forward these major initiatives in the field of ensuring international peace and security:

- Creation of NuclearFree Zone in Central Asia;
- Reconstruction Afghanistan;
- Creation of the International Centre for Struggle Against Terrorism;
- Cooperation among UN and Central Asian countries for regional security and ecological problems (examples: Aral Sea, water problems);
- Central Asia in the context of millennium development goals;
- Cooperation among Central Asian countries and UN agencies, including UNESCO.

The United Nations Department for Disarmament Affairs actively participated in development and financing the contract draft on the Nuclear-Free

Zone in Central Asia, which is planned to be signed this year.

Partnership relations concerning issues of safety in Central Asia

In Central Asia, the matters of safety, peace and stable development that while coming into world society was Uzbekistan President Islam Karimov's first step in 1993 at Session 48, and in 1995 in Session 50 he spoke of his concerns.

President of Uzbekistan Islam Karimov made a speech and told the world of the difficult matters in Central Asia. Nearly 100 billion people live in Central Asia, including different ethnographic people and religious groups. In recent years, this region collected nuclear weapons and ordinary weapons. Many crashes happened in that region. Eighty years of the twentieth century were filled with economic and moral tightness. For example, in Tajikistan there was Civil War; Afghanistan was more active, deeply economic and public matters happened. It is important to save this country from nuclear weapons, to stop wars in Tajikistan and Afghanistan. This requires cooperation with large companies, including UNO, to address major concerns: religious extremism and fundamentalism, international terrorism, narcobusiness and narcotographic and ecological problems. That's why, including head matters were making between countries of Central Asia and UNO.

International terrorism is a global threat to peace and safety of people. This illness knows no limit, and terrorism is living in Central Asia. The problem of Central Asia at that time was active international terrorism. That's why they needed a source of people. In consequence, developing international terrorism in Central Asia depended on inside and outside matters. In 2000, a seminar took place against international terrorism. In September of 2000, the previous secretary of the UNO, Kofi Annan, considered this problem. President Islam Karimov took international terrorisminto his consideration and admitted Uzbekistan was ready solve the problem [1,109]. International terrorism aof Central Asia, but the UNO had many chances and organizations for settling this matter.

Cooperation about the Afghanistan problem

The most important matter of UNO is, without a doubt, the problem of Afghanistan. Afghanistan was a point of fire with the people of the region. The War in Afghanistan was not only Central Asia; the problem of Afghanistan was a global safety issue.

This problem hasn't been solved positively yet. The first time President Islam Karimov spoke about Afghanistan's problem was in 1993 on 28 September at Session 48 of the General Assembly of the Untitled Nations Organization. Two years later, in 1995 on 24 October, President Karimov announced his ideas at Session 50 of the Untitled Nations Organization and he showed his initiatives, outlined below.

Firstly, solve the acuity in Afghanistan, at first limiting interference of foreign policy.

Secondly, weapons mustn't be taken to the territory of Afghanistan. Afghanistan was not a territory of war; it must have peace and steadiness, which may help this country in developing economical-social life [2, 57]. According to the strategy, Afghanistan is situated conveniently. It had the possibility to go out to the sea ports. Karimov's invitations and initiatives were approved by UNO, especially that weapons mustn't be taken to Afghanistan. In 1996, this matter was supported by the Security Council of UNO. The events of Afghanistan were discussed by the Security Council of UNO twice in 1996, on 15 February and 22 October. The countries of Central Asia were supporters of peace. That's why these countries took Afghanistan into their consideration. For example, in 1997, Uzbekistan arranged a group talk with six countries bordering Afghanistan, plus Russian and USA. Afterwards, this group was called " 6+2." It included the six borders of Afghanistan: China, Pakistan, Iran, Uzbekistan, Turkmenistan, Tajikistan, plus Russia and USA.

In 1997 on 16 October, the first meeting of "6+2" took place in New York about the Afghanistan matter. In 1999 on 14 January, the next meeting of "6+2" met in Tashkent. The biggest achievement of "6+2" in Tashkent was that two main groups in Afghanistan that is to say motion of "Taliban" and it against United Battle Afghanistan were sitting together.

At the end of the Tashkent meeting, attendees signed the "Tashkent Declaration" about "Solving the problem of Afghanistan by peace." NATO accepted it in 2008 on 2 April and President Karimov made a speech and talked about the Afghanistan matter. The most important matter was setting up the "6+2" that was active during 1997- 2001 years. On NATO, the President of Uzbekistan Islam Karimov made a speech and he discussed: "He would like to say the "6+2" turned into "6+3" [3, 317]. The problem of Afghanistan wasn't a regional problem; it was a global problem. Around the world, many meetings, international conferences, seminars were organized, but not all of them could be effective. That's why Islam Karimov again and again took into his consideration the Afghanistan problem. In 2010 on 20 September in the General Assembly of UNO, Islam Karimov said: "Afghanistan was the main duty for us." One more time he took into consideration the "6+3." The result of I. Karimov's initiatives and ideas was catching the world's interest. As such, after taking out the force of ISAF from Afghanistan, Uzbekistan will emphasize what happened in Afghanistan and affects the countries of that region.

The connection of partnership because of Aral Sea

One of the global problems is the Aral Sea in Central Asia. The trouble of Aral was the half of XX century and XXI century. In Session 48 of the UNO, Uzbekistan President Karimov appealed to countries: "The trouble of Aral Sea caused many misfortunes. We must appeal to the world society. We called them for saving Aral and near the Aral." Islam Karimov emphasized the problem of Aral, and he initiated the organization of the special commission about the Aral Sea.

After that, in January 1994, the meeting of the heads of the republics in Central Asia was organized in Nukus about the problem of the Aral Sea. In this meeting three-five year's certain operations were formed and stated about improving the ecological state of the Aral and developing socio-economic of the zone. In March 1994, the third meeting was organized about this problem in Toshhovuz. In February 1997, an important meeting of the heads of five republics in Central Asia was formed with the delegation of the UNO, the World Bank and other international organizations. Practical trainings were formed for developing a function of the International Fund for rescuing the Aral Sea, which was organized in the 1993 meeting in Kyzylorda.

"Declaration about the problem of developing the Aral Sea of the states of Central Asia and the International organizations," [4,128] which was organized about settling the problem of the Aral Sea on 20 September 1995 in Nukus, has important and global significance. In fact, destruction of the Aral Sea is a very important problem in Central Asia because about 50 million people live near the Aral. In the 1960s, flowing of the Amu Darya and Sir Darya to the Aral was 90%, exactly 60 km^3 flowing of total water volume. In 1992–2000, this result decreased to 8.5 km^3 [5, 12]. According to suppositions of UNESCO, in 2000–2005, a year's flowing to the Aral Sea decreased again to 5.2 km^3. The problem of the Aral Sea is a global one, which concerns the whole planet. For this reason, cooperation of the UNO and the countries of Central Asia has international importance. On 11–12 March 2008, an international conference was formed about "The problem of the Aral Sea, its genofond, effect of fauna and flora, and international measures about facilitating its results," with Uzbekistan and active delegations of the UNO in Tashkent. In that conference, I. Karimov, the President of Uzbekistan, spoke about implemented measures of settling the problem of the Aral Sea [6, 287]. That conference accepted the Declaration. In April 2010, Ban Ki-moon, the Secretary General of the UNO, visited Uzbekistan. First of all, in Nukus, the Secretary General of the UNO took part in the presentation about implemented measures of improving the ecological state of the Aral. Ban Ki-moon said that he's ready to settle results of the ecological destruction, and to solve the problem of the Aral Sea [7, 1].

The Aral Sea was a very beautiful sea; it had its rare flora and fauna in its wonderful zone, but now it is deteriorated. Being sovereign of Russia to cotton in Uzbekistan being careless to ecological problems were the cause for this destruction. For forty years, the water area of the Aral Sea shortened seven times, water volume decreased 13 times, its mineralizing increased ten times and it was worthless for alert, living organisms in the sea. In consequence, nearly all animals were lost. Today, the Aral Sea has ecological problems and also socio-economic and demographic problems.

The problem of the Aral Sea is a problem for millions of people who live there and who ask help with great hope from influential organization such as the UNO. The international fund for rescuing the Aral Sea was given the level of observer in the UNGA, among the cooperation of UNO and the countries of Central Asia, which was ratified by all country members in Session 63 of the UNGA.

Conclusion

In the end of XX century, when abolishing the previous Union and the finishing of the "Cold War," new countries appeared in Central Asia .

In Central Asia at that time, ecological, economic, military, transport, ecology problems appeared. To solve the troubles, Central Asia and the UNO cooperated together. We may think of the following ideas in summary:

- The cooperation between Central Asia and the UNO was of a high degree. The main direction between the countries of Central Asia and the UNO was social-ecological and political matters, peace in the regions, safety and steady development.
- In Central Asia, all regional and global problems were indicated among the UNO to the World Society. And for solving the problems, they decided to work together. These problems were:
- The problem of Afghanistan,
- The problem of the Aral Sea and its ecological issues, and
- The initiative on creating the Nuclear-Free Zone in Central Asia.

For more successful implementation of the integration processes of interstate cooperation in Central Asia, it is necessary to accomplish large-scale and long-term regional projects in various areas of economic, communication, and humanitarian cooperation, which will allow for the development of cooperation with various regions of the world, and certainly for strengthening the security of Central Asia.

To generalize and come to one united conclusion, efficient cooperation of Central Asiatic states with UNO and other international organizations can successfully serve to reform the social, economic and political sphere, as well as vastly improve global and regional safety and stable development.

References:

1. Balance of convenience / Constitutional and Public Law. – M. – 640 p. (in Russian)
2. Karimov, I.A. Vatansazhdagokhkabimukaddasdir. – Ch.3 – Tashkent: Uzbekistan, 1996 – 366 p. (in Uzbek)
3. Mayers, D. Social Psychology / Transl. from English, Saint Petersburg, 1999. – 710 p. (in Russian)
4. Kholbekov, A.Zh.,Matiboev, T.B. Izhtimoiyadolatvademokratia: barkarortarakkieyotiyulida. – Tashkent: "Yangiasravlodi" Publishing House, 2004. – 210 p. (in Uzbek)
5. Khayrullaev, M. Izhtimoiykhayotvamafkura //Uzbekistonningmilliyistiklolmafkurasi. – Tashkent, Uzbekistan, 1993. – 180 p. (in Uzbek)
6. Matibaev, T.B. Uzbekistan fukarolikzhamiyatiasoslarinibarpoatishzharayonidaizhtimoiyadolatvademokratiyaninguzaroalokadorligi. Social fanlarinomzodi – Dissertation autoabstract – Tashkent: 2008 – 172 p. (in Uzbek)
7. Uzbekistan Republication Constitution khukuki/Muallifarzhamoasi. Masulmukharrir, PhD in Law, Professor A.Kh.Saidov – Tashkent: "Molia" Publishing House, 2002. – 410 p. (in Uzbek)

The Standards of Coordinating the Benefits of the Centre and the Provinces in Uzbekistan

Ruzimat Juraev
Namangan State University, Namangan, Uzbekistan

Although the process of coordinating the benefits of the centre and the provinces in Uzbekistan hasn't been sufficiently studied, the justice, equality, legal, balanced and the other requirements, principles and the standards of coordinating the benefits of the centre and the provinces have been recommended by the scholars [1, p. 56]. The unsystematizedness of these requirements and not including all the aspects of this process demand to make them general and broad. In our opinion, it seems more logical to express the requirements of coordinating the sate-wide and regional benefits in the norms of justice, efficiency and stability. To follow these norms serve to gain the intended aim, that is, to develop the mechanism of realizing the benefits of human, to optimize the political administration.

As the president Islam Karimov mentioned, we are striving for building not just a simple democratic state, but a fair democratic state. Striving for justice is an important peculiarity of our nation's moral-spiritual world [2, p. 355]. In gaining the coordination of the state-wide and regional benefits justice is considered to be as the basic standard [3, p. 643].

The progress of the ideas of social justice in Uzbekistan was deeply analyzed, and their coincidence with democracy was proved by A. Kholbekov and T. Matiboev. Agreeing with their opinion, we can say that being disposed towards this or that aspect in democracy can stay firm. For example, there can be difference between freedom, equality, everyone's participation in the political processes, majority's government and protecting minority, legal bases of the state and the social direction of the state. Only the rules of political justice can give the right and wise solution in this case, that is, it provides the balance [4, p. 43]. And because the rules of political justice are expressed in the political and legal norms, the process of coordinating statewide and regional benefits demands to reflect them objectively.

Indeed, on the way to strengthen the independence, carry out the aim of building up a new society, it is not possible to realize it without leading and organizing the people towards it. And it requires creating a new idea leading and teaching the people towards this aim. Every society must have its own ideas protecting its aims, organizing and conscripting the people around this aim. Such social ideas appear as the result and a certain adaptation of spirituality [5, p. 31]. The adaptation of national ideology of the Republic of Uzbekistan and the firmness of the law reflects in putting the ideas of constitutionalism into action in our country.

As T. Matiboev mentioned, "In the citizenship society the factor leading the people towards the necessary activity is the appropriateness of benefits. If the adaptation of social relations provides the coincidence of benefits, its reflection will consist of social conflicts. The main mechanism in preventing the increase of conflicts in the social relations is the social justice standards, because, the laws of justice put on the basis of law, freedom and the equality make everybody obey the rules of democracy" [6, p. 22–23]. In addition, it is not possible to present the social justice and democracy in the life of the society without the political activity and participation of the people, and in its turn, it requires to fully follow the principles such as, openness, awareness, variety of ideas, multiparty, criticism and creativeness [6, p. 8].

It is known that the natural settlement and closeness of geographical regional borders, natural features and economic conditions of the region, its industry and agriculture, the settlement of the enterprises and the orgnisations and their specialization, the existence of the means of transport and communication are the factors of administrative-regional division [7, p. 360–361]. The economic basis is one of the main requirements to put the local government in force. Though the economic basis of the local state governments has already been entirely streamlined, developing such foundations of citizens' self-governing authorities is the urgent question. Because, one of the aims of coordinating the state-wide and regional benefits is to provide their managing the economy independently.

Generalizing the experience of foreign countries, it can be said that the achievement of coordinating the state-wide and regional benefits requires to make a convenient environment to set up new economic relations and general economic conditions making easy the work of economic subjects with different property in the regions, to provide conditions for realizing the regional benefits; to help the local self-governing, to restore the country's single economic site, regional division and integration of social labour, to restore the international infrastructure serving to develop the market relations, to help the regions which are economically behind.

References

1. Баланс интересов // Конституционное и муниципальное право. — М., — 640 с
2. Каримов И. А. Ватан саждагоҳ каби муқаддасдир. Т. 3. — Т.: Ўзбекистон, 1996. — 366 б.
3. Майерс. Д. Социальная психология / Пер. с англ. СПб., 1999. — 710 с.
4. Холбеков А. Ж., Матибоев Т. Б. Ижтимоий адолат ва демократия: барқарор тараққиёт йўлида. — Т.: "Янги аср авлоди" нашриёти, 2004. — 210 б.
5. Хайруллаев М. Ижтимоий ҳаёт ва мафкура // Ўзбекистоннинг миллий истиқлол мафкураси. — Т.: Ўзбекистон, 1993. — 180 б.
6. Матибаев. Т.Б. Ўзбекистонда фуқаролик жамияти асосларини барпо этиш жараёнида ижтимоий адолат ва демократиянинг ўзаро алоқадорлиги. Соц. фанлари номзоди… диссертация авореферати. — Т.: 2008. — 172 б.
7. Ўзбекистон Республикасининг Конституциявий ҳуқуқи / Муаллифлар жамоаси. Масъул мухаррир ю. ф.д., проф. А. Х. Саидов. — Т.: "Молия" нашриёти, 2002. — 410 б.

Kazakh Enlighteners' Views on Counseling Political and Social Problems

Murat O. Nassimov[1]
Bauyrzhan Akhmetov[2]
Botagoz Paridinova[1]

[1]«Bolashak» University, Kyzylorda, Kazakhstan
[2]Korkyt Ata Kyzylorda State University, Kyzylorda, Kazakhstan

Abstract. The Kazakh thinkers sought to understand the world, existence and way of life of their time deeply. Kazakh philosophical views taken the source from the real life are full of thoughts on society and nature. So, such socially related approaches are of consulting features. Social and political counseling means the use of consulting technology elements.

Keywords: Social and political counseling, Kazakh enlighteners, Abai's wise words.

1. General concept interpretation

Counseling, or as we like to call it in today's society Consulting occupies a special position in the social, political, and cultural and social structure. Every day a person is associated with overcoming a variety of organizational, psychological, moral, and other difficulties. Therefore, the life of modern humanity depends on many factors and is not at all simple.

The word «advice» from the English language and means counseling, pay counseling. At the end of the twentieth century, the concept came into use modern Kazakhstan, but it is also used as advice. On the issue of consulting author has devoted some research in kazakh language [1].

Dictionary of Russian S.I. Ozhegova a definition was given: «1. Meeting of experts case or subject. 2. Advice given by a specialist» [2].

The interpretation of consulting notion is counseling. It is an intellectual action in solving complex problems of any phenomena of organization development and management. Above mentioned activity is of great significance. According to some scientists, the first consultant was the Chinese philosopher Confucius well-known by his works. In ancient times philosophers, great speakers and storytellers played a great role in political and social counseling. Therefore, we are fully confident that the speakers and over lings are the sources of political consulting assessment in Kazakhstan. Including thinkers and enlightener scientists' works are of great share in forming people's political and social views.

In the history of Kazakh social beliefs enlightenment literary works are of great importance. In the nineteenth volume of twenty-volume work, on

Kazakh great literary scholar, playwright and public figure M. Auezov's opinion: «If we recall the past history of the Kazakh SSR, we can proudly name the first three figures of Education. Although being natives of distant suburbs of huge Kazakhstan they seem close to each other and noble. They are the greatest men, Shokan, Ibrai, Abai.

In spite of having different ways of life and creative work they all three have obvious similarities in living in the same Motherland. Although they were born in the Kazakh yurt and were representatives of their native land, they were proud of their national language, art and historical culture, and all of them were educated in the Russian environment. They were the first Kazakhs stepped on the land of great knowledge. So, they got an excellent education.

If Shokan hadn't mastered leading democratic Russian culture, he wouldn't have become Shokan. If Ibrai hadn't got acquainted with the Russian culture and if he hadn't known Ushinsky's pedagogical approaches and if he hadn't learned Russian humane and progressive classical literature, he wouldn't have become Ibrai. If Abai hadn't learned and mastered literary heritage of Krylov, Pushkin, Lermontov, Saltykov-Schedrin, he wouldn't have become Abai» [3].

For this reason, we intend to consider political and social counseling in the works of Shokan, Ibrai and Abai in this article.

2. Polititcal and social views of Shokan, Ibrai and Abai

The Kazakh thinker, ethnograph, historian, geographer, outstanding scientist and enlightener Shokan Ualikhanov's scientific researches and social views are still of great value. His articles «Abylai», «Some features of Shaman religion among the Kazakh people», «Notes of court reform within Siberia», "Dzungarian sketches" describe establishment and development of our nation.

For instance, Shokan overviews the life of the famous khan in his article «Abylai». He stated that those periods were the age of heroism for the Kazakh people. He describes his way of the leader and how he was in captivity, and at last it was clear that he had features of a forecasting politician.

In article «Some features of Shaman religion among the Kazakh people» the scholar presents scientific approaches to the features of Shaman religion among Kazakhs. It was defined that people kept Islam religion including Shaman traditions and superstitions. Therefore, ancestors' spirit, ghosts, orgones, Shaman's methods of treatment and venerating shrines, sacrificing livestock, Kazakh cosmic notion, the Holy spirit of nature are mentioned. Shokan suggests «The fairy-tale about dead and a living soul and their friendship» in this article. He defines the essence of friendship between dead and alive fellows. He eulogizes to keep traditions, to hear somebody's advice and strive for being brave by this saying: «*My son, keep in your mind. If you were left alone at abandoned place, pray to God. If you were on the edge of the grave at night stay overnight there*» [4].

Ibray Altinsarin, the first innovative teacher, great pedagogue, poet, writer, played a great role in literacy by directing European standard education to the Kazakh schools and art. Ibray tried to call the youth to seek for knowledge and art not by insignificant words, he tried to prove his words by exactness. Students understand the new theme easily thanks to his different methods of teaching. First of all, he uses oratorical skills successfully. In his each verse he says:

«Children, let's study by praying to God and to learn a reading material well». Ibray finds the way to each child's heart by repeating these verses again and again. He compares light and darkness in order to differentiate scholarship and ignorance. So, he equates light to study:

«Children, if you study it lightens a lamp, and you can find everything without looking for it.» He describes ignorance as: «Someone, he doesn't study, will stay in darkness».

The author pays a great attention to the announced knowledge that gained by labor and search. He notifies that it needs tolerance and evenness. And also he says searching foe the knowledge is everlasting and waste less.

We can say that Ibray's poem «The people have the knowledge» is of new content and feature. The poet describes technical devices and equipment, their movement and work here; they serve not only by the conception of the Kazakh children, but adults as well. It is such technical …radio, telegraph, telephone, gas, power, plane etc. was fantastic for the children. And to explain all these for the Kazakh children he uses the methods of puzzles. It is not necessary to prove that, but some features of this method will be mentioned.

The author tries to teach them by using comparison of the unknown techniques with the things we have in real life. For example, he images the locomotive as «*a cart without horse*». And he compares a ship sailing in the water *with a large fish*. And it is really clear for the children who accustomed to recognize the meaning of the thing by guessing the riddle. Therefore, all these poems are considered to be of incomparable cognitive features.

I. Altynsarin was the first founder of prose genre in literature. He counsels the schoolchildren by honesty, trust, honest work and justice, decency and calm, wise and education in his literature. The writer pays attention to the advantage of settled way of life in his stories as «Kypshak Seitkul», «Kiiz ui and agash ui» (yurt and wooden house).

The writer's stories, «Compassion more than sickness», «Noble (Fine) grass», «To benefit from livestock», are devoted to teach children to such ideal features as charity and sympathy, calm and restraint. Ibray is recognizing the children's notion paying attention to form the event briefly, as a fable and a riddle. He pays attention to develop their thinking abilities by these works.

We can single out «Sons of poor and rich» among his works. In this work the author pays attention to the great social problem using the main characters Asan as the representative of the rich and Usen is the son of poor. He made conclusion by representing the influence of environment and the family to the children's upbringing. He describes experience differences of these two boys as they were brought up differently, as for Asan who was grown up in the rich family and had no life experience and opposite to him Usen, the son of the poor, hard-working and experienced arrived at a solution in difficult situation. And he suggests an idea that labor and experience will win at any complicated situations [3, pp. 8-17].

The Kazakh great poet, philosopher, the founder of written literature Abai's political and social, anthropocentric, humanistic views are of great significance. His each poem, wise words, verses and translations uncover the past, conditions of life, fortune and people's dream. And we can clarify our past, future and present by understanding the main idea of his works deeply.

In his first wise word Abai says: «How will I spend my future life?» and asks by «nation», «cattle», «science», «religion», «child», «Will I take care of ?» At last he says: «I will amuse the white paper and black ink, if somebody finds a required word, he will write and read…» [5]. Let us take into consideration the consulting features of his wise words.

The third wise word says: «*One was elected as a bolys for three years. The first year is spent by people. You were elected by us, weren't you?*» The second year is spent by snooping around candidates. The third year is spent by looking forward for the next election…. The poet's description of those days democratic formation values reminder us the image of nowadays election campaign. The technology of electing bolys shows us that to be well educated was the precondition on those days.

It is said in the eighth wise word: «*Who will learn this recommendation and who will listen to propaganda? One is bolys, one is bii. If they were eager to learn this recommendation, listen to propaganda and he wouldn't be elected to this post …*». The poet states morbidly that the both, bolys and bii, are limited by their thoughts of being the best one, elected to be a model and to give an advice; devotion, humanity, wise, science, education are all inexpensive. Abai isn't satisfied with their appreciating social development basic values that influences the poor's opinion not to be in need of education, science and advice. Otherwise, it is not a secret that nowadays such views are still in people's mind.

Abai's such ideas of differentiating wise person from unwise (*15th wise word*), competition of power, mind and heart and their admiration for science (*17th wise word*) prove the altitude of wisdom and science. It shows that science is unusual function of human being in systemizing objective sides of environment. The basis of this function are getting information, systematizing them, examining and synthesizing a new doctrine. Science does not only describe but also helps to make prognosis about controlling nature and social phenomena. So, we notice the model of making prognosis in Abai's creative work.

There by assessment was made to political and social opinions of these three figures in the Kazakh policy. It was noticed that Kazakh genii call members of society to peace, kindness, to be literate and advice to follow science. That's why, it is obvious that Kazakh genii's opinions won't loose its actuality and value.

References:

1. Nassimov M.O., Ensepov A.A., Paridinova B.Zh. Political Consulting: Main Problems // Молодой ученый. – 2013. - №11.1. – С. 25.
2. Ozhegov S.I. Russian dictionary: 70,000 words / Ed. N.Y. Shvedova. - 21th ed. rev. and add. - M., 1989. - P. 292.
3. Zharmukhamedov M. The studies of I. Altinsarin. - Almaty: Zhazushy, 1991. Pp. 20-21.
4. Ualikhanov Sh. Articles and letters. – Almaty: the Kazakh united state publishing house, 1949. – P. 169.
5. Kunanbaiuly A. Two volumes of complete collected articles of his works. – II volume. – Almaty: Zhazushy, 1995. – P. 158.

International Terrorism: Definition, Essence, Main Features

Alexander Pokhilko

Pyatigorsk State Linguistic University, Pyatigorsk, Russia

Abstract. *The article covers main features of international terrorism as one of the global problems of modern times, forms of its expression, strategy of terroristic activities.*

Keywords: *international terrorism, terrorist organizations, bipolar world, social-political sphere, public relations, state structure, international terrorist centers.*

The problem of international terrorism is one of the global problems of modern times. There are several reasons: 1. International terrorism is widely spread not only in the regions of traditional international conflicts, such as Middle East or South Asia, but also in developed states of Western Europe and the USA; 2. Every year many acts of international terrorism are performed in the world, resulting in thousands of killed and injured people; 3. Overcoming such global problem as international terrorism requires total efforts of many states and nations of our planet, all world community; 4. It is obvious that the problem of international terrorism shall be reviewed as part of the whole complex of global problems.

The purpose of our article is to define the essence and main features of international terrorism and their characteristic.

Gusher A. I. noted: «Terrorism in any form of its expression turned out to be one of the dangerous by its extent, unpredictability and consequences social-political and moral problems, which the mankind enters the 21st century with. Terrorism and extremism in any forms of its expression more than ever threaten the safety of many countries and their people, involve large political, economical and moral losses, subject large quantities of people to psychological stress, the further it goes, the more lives of innocent people it takes» [2]

The gravity of international terrorism problem is highlighted by researcher Panin V. N., who pays attention to the fact that terrorism «has become one of the most critical safety problems. As alarming tendencies of recent years there should be noted the desire of terrorist organization leaders to make their actions global, to present their actions as the conflict of Islamic and Christian world, struggle for freedom and independence and other grounds of their actions attractive for part of world community and extremist forces. In modern world one can see information, tactical, resource support of terrorist organizations in separate country as well as worldwide» [4, page 57].

Activation of international terrorism, as stated by Maruev A. Yu., «is the result of profound changes in the system of international relations. This situation occurred mostly due to the destruction of international safety system of the "Cold War" period that was characterized by bipolar balance of forces. This system restrained the number of main forces acting on the world stage, thus facilitating the management of global problems.

The destruction of bipolar world and existing system of international relations revealed a row of problems. The so-called "Third world" that was shared by two superpowers and controlled by them got beyond the control and began to play its own game, choosing methods of violence for reaching the goals with increasing frequency. Vacuum of force in the system of international relations was quickly filled with different extremist and other destructive forces. It was expressed in surge of terrorism that covered the whole world and threatens the safety of many states» [3].

The issue of definition "international terrorism" as well as "terrorism" remains pending, there is no universal definition. Philosophers, political experts, historians, psychologists, legal experts, sociologists, linguists, etc. study terrorism regarding the objects of their researches, but all of them define the same features that are peculiar for international terrorism. Though, as social-political occurrence, international terrorism reflects conflict interaction of different forces that is based upon, as a rule, struggle for power or possession of material or spiritual values.

Political experts note that terrorism was never developing continuously or sequentially, on the contrary, it appeared there and then, where and when there was a ground favorable for its occurrence. Considering global experience it is clear that eruptions of terrorism happened at those historical periods that were characterized by the escalation of contradictions in social-political sphere, breaking of public relations, changing of state structure, lack of stability, acts of nationalism and separatism, spread of criminality.

The main features of international terrorism are the following: global demonstration, severe and negative dynamics, need of immediate solution, etc. But together with this international terrorism has its specific features as well. International terrorism is closely related to the main spheres of world community: national relationships, politics, religion, ecology and etc.

Modern burst of international terrorism represents the expression of acute contradictions caused by uneven development of countries in the world. This phenomenon is a specific reaction against "abjection and humiliation by established world order, one-side hegemony, reaction against the efforts to monopolize the right to apply legitimate violence and implant foreign values" [3].

According to A. I. Gusher's opinion the activity of international terrorism in modern conditions is characterized by wide spreading, lack of expressed state borders, connection and interaction with international terrorist centers and organizations; fixed organizational structure that consists of management and operational level, intelligence and counterintelligence units, material and logistic assistance, battle groups and cover; strict conspiracy and careful selection of staff; presence of agents in law-enforcement and state authorities; good equipment competing and sometimes surpassing equipment of government troops; presence of extensive network of secret hideouts, training facilities and landfills.

It is characteristic that when international terrorism gets modern means of conducting information war in its hands, it imposes its ideas and evaluation of the situation upon people, widely and successfully settles mobilization tasks on attracting young people, say nothing about professional contractors.

International terrorism, as well as terrorism in general, can be expressed in different forms, such as political, nationalistic, religious, criminal and ecological.

Political terrorism is characterized by the fact that the participants performing political terror pursue the aim to do political, social or economical changes inside this or that state, and together with this break state-to-state relations and law enforcement.

Nationalistic (ethnic or separatist) terrorism has the aim to settle national issues in different multi-ethnic states.

Religious terrorism is an attempt of armed groups belonging to a particular religious denomination to fight against the state, where the other religion or other religious trend is practiced.

Crime terrorism is based on some criminal business (drug traffic, illicit trade in arms, drugs, smuggling, etc.). The purpose of this variety of international terrorism is creating tension and most likely obtaining extra profits in these circumstances.

Environmental terrorism is a kind of terrorism implemented by groups that use violent methods opposing scientific and technological progress, environmental pollution, killing of animals, construction of nuclear facilities [1, page 35–36].

The number of terrorist organizations is growing, the level of their organization is increasing, the interaction between the individual terrorist groups is strengthening, their efforts are united in conducting large-scale operations. In some cases there is the formation of some kind of "terrorist international" directing its efforts to creation of a sort of "terrorist enclaves" at the territory of states, where for one reason or another there was a vacuum of legitimate political authority (Afghanistan, Chechnya, Iraq, Kosovo).

Terrorist groups have an extensive network and coordinate their actions. Terrorist acts in most cases began to achieve their goals that are not only causing direct harm to victims, but also the implementation of terrifying effect: as a result of these acts they manage to instill fear, cause feeling of confusion, helplessness, endanger wide range of people.

Financial, economic and technical capabilities of terrorists increased abruptly. Some terrorist organizations have the potential comparable in terms of volume to the military capabilities of small states. The continuing increase of terrorism financing amount is extremely dangerous. Experts found the following pattern: the more efforts public authorities make for fighting against terrorism, the greater the amount of financial assistance to extremists from "terrorist international", mafia communities and all sorts of "charitable" funds created by them and other financial institutions are.

It should be noted that the main feature of international terrorism is its unpredictability. Very often the subjects of terrorism are very ambitious politicians, mentally unbalanced people, who see terrorism as the only way to achieve the objectives on the world stage and in international relations.

The strategy of terrorist activity is in skillful use of small resources directed against the opponent that is greatly superior in power, potential and capabilities. Herein is the main difference between international terrorism from other forms of conflict.

With the development of scientific and technological progress means and methods of terrorist exposure continue to improve and become more sophisticated: there is a continuous search for new ways and means of influence. A real threat became the possibility of application of radiological, chemical or biological weapons by terrorists. Terrorists actively use the Internet. The specific character of global network allows easy access regardless of geographic location, unlimited potential audience, high speed of data transfer and together with these the difficulty of control by law enforcement agencies.

One of the main tasks performed by extremists via the Internet is the wide coverage of terrorist attacks with their reference to extremists' ideological messages and terrorizing the society. Internet is an effective means of influencing the population. Mass propaganda is one of the main activities of terrorists in the Internet, which is used to recruit new members, including suicide bombers from Islamists and extremist-minded young people, to mobilize supporters playing an active role in supporting terrorist organizations. Internet is used as a means of psychological warfare and as a means of disinformation, intimidation and manipulation of public opinion for the purpose of substitution of concepts and facts.

Also in the process of terrorist attacks preparation the Internet is used by terrorist organizations to collect and analyze data about possible targets, information about tactics and means of upcoming attacks using powerful search systems, using open and confidential discussion groups.

A large role in the struggle against international terrorism is played by mass media, which should provide efficient information countering the propaganda activity of terrorist organizations and groups seeking to plant fear among the population and break the will.

As Maruev A.Yu. thinks, "in today's geopolitical environment international terrorism is directly involved in the global geopolitical confrontation. The organizers of terrorist activities as well as some of their opponents referring to the fact that the majority of terrorist attacks in recent years were made by the representatives of Islamic extremist organizations try to interpret the events of recent years on the world stage as the "conflict of civilizations". More often it is presented as the conflict between Islamic and Christian civilizations. It is stated that in the geopolitical space there is the so-called war between the "civilized" North and the "barbaric" South [3].

Thus, based on the above, it must be concluded that the growing problem of international terrorism as well as other problems of the present can lead to death of all mankind. The problem of international terrorism can not be solved in one plane and one way. It is necessary to achieve understanding among the general population of all countries of the complexity and fragility of the modern world. It is necessary to identify the sources of tension, where the intent to use terror for political solutions and other tasks is increasing, where people capable to carry out attacks are brought up.

Religious education of good quality is becoming very important today, because terror ideology must be eliminated via hard spiritual work.

Terrorism in all its forms must be condemned by the international community, regardless of the purposes proclaimed by the terrorists and those arguments that they bring to justify their actions.

References

1. Grachev S. I., Kolobov O. A., Kornilov A. A. The United States of America and International Terrorism. Nizhny Novgorod, 1999, page 35–36
2. Gusher A. I. Problem of Terrorism at the Turn of New Millenium of Mankind New Era // www/silovik.net
3. Maruev A. International terrorism: Mankind Future is Questionable // Red Star, 2007, 17–23 October // www/klaipeda1945.org
4. Panin V. N. Religious extremism as one of the factors of destabilization of social-political life at the North Caucasus //Islam at the South of Russia. Issues of revival and development. Materials of International scientific and practical conference. — Pyatigorsk, PSLU. 2008. — page 57

PSYCHOLOGY

The Impact of Meditation on Emotional Intelligence of Migrants as a Key Factor of Social-Psychological Adaptation

Sergey V. Afanasyev
Moscow, Russia

Introduction

Migration problems are particularly acute on the agenda of public debate in the context of globalization. Migration in the context of migrants' social-psychological adaptation requires special attention as unresolved psychological problems arising in the process of migration have result in threats, both for migrants, and for a society in general. People contacting with a new culture can objectively consider this situation as complicated and strained, especially at an initial stage. Stress levels occurring on the background of cultural and religious differences can be particularly high when the culture of the country of origin is significantly different from that of the host country. On the one hand migrants experience a severe need in successful and effective socialization in a new society and culture. On the other hand a need in psychological health maintenance is important since the mentality and emotions are tensed during the migration.

In connection with the foregoing, the problem of migrants' social-psychological adaptation seems to be very relevant in cultural, social, psychological aspects which in turn affect the business and public sectors. The research subject is more significant because of a lack of researches relating to the impact of meditation on the emotional intelligence as a key factor of the migrants' social-psychological adaptation. The research aim is to investigate a role of meditation in the emotional intelligence development as a key factor of the migrants' social-psychological adaptation.

The general hypothesis of the research is the assumption of dependence between meditation practice and dynamics of migrants' emotional intelligence as a key factor of their social-psychological adaptation. On the basis of the general assumption the author formulated the following hypotheses:

1. Meditation affects dynamics of the emotional intelligence.
2. Dynamics of emotional intelligence that arises due to the practice of meditation is positive.
3. The practice of meditation enhances key intrapersonal and interpersonal competencies of emotional intelligence causing the efficiency of social and psychological adaptation of migrants.
4. The choice of a specific method of meditation does not affect significantly on the dynamics of the emotional intelligence.

The analysis of published studies shows that the success of the integration process of acculturation through dialogical mode of communication is largely driven by certain participants' certain set of intrapersonal and interpersonal skills designated as intercultural competence in models of emotional intelligence. Intercultural competency is defined as the ability to communicate effectively in cross-cultural situations and to relate appropriately in a variety of cultural contexts (Bennett and Bennett, 2004). Besides, the intercultural competence is to help to the effective interaction in the foreign environment not only at professional, business level, but also at an interpersonal, emotional level. Also in addition to increased efficiency of the cross-cultural communication, the optimization of interpersonal relations and socially-psychological adaptation should be related to the major functions of the emotional intelligence (Pankova, 2011). It is important to understand that a skill of the effective communication on the cultural borders, an emotional balance and achievement of the creativity within the collective behavior becomes one of the most significant abilities in the today's world, including in aspect of the migrants' social-psychological adaptation, as a social group which first of all requires for searching for effective tools of impact on emotional sphere within the migration process. In connection with the aforesaid, studying of potential factors of the management and impact on the emotional intelligence as the factor of the migrants' social-psychological adaptation is very important taking into account the method optimization of estimation of the emotional intelligence dynamics.

Method

The theoretic and methodological basis of the research is presented by theoretical provisions and conceptual models of the emotional intelligence (R. Bar-On, P. Salovey, D. R. Caruso, J. D. Mayer, D. Goleman, D. V. Lyusin), modern and classical concepts of the social, plural and cognitive intelligence (D. Wechsler, E. Thorndike, S. Stein, H. Gardner, W. Payne), actual theoretical and empirical substantiations of the meditative practice efficiency (J. Kabat-Zinn, Maharishi Mahesh Yogi, J. Teasdale, Dharma Singh Khalsa, A. Weintraub, N. Rosenthal), and key provisions of the socially-psychological approach to research of an individual's adaptation (R. Lazarus, S. Folkman, J. Zhang, M. Goodson, V.S. Ageiev, V.N. Miasishchev, A.A. Nalchadjan, T. G. Stefanenko).

The selection of the research methods was mainly due to the theoretical and methodological basis, as well as to the hypothesis of the study. The main research methods were Thematic Analysis, Interpretive Phenomenological Analysis, and interview as qualitative research methods, as well as a quantitative method of estimating the level of emotional intelligence "EmIn Questionnaire" by D.V. Lyusin. The data received during the research were statistically processed in SPSS Statistics 17.0.

The main methodological problems of emotional intelligence researches should include the selection of adequate and valid methods of its measuring and evaluation. As noted above, the present study attempted to study the impact of meditation on emotional intelligence as a factor of successful social-psychological adaptation of migrants by integrating qualitative and quantitative research methods. This research focuses on the understanding of personal subjective experiences of those who practice meditation in the context of its impact on emotional intelligence as a key factor of the migrants' social-psychological adaptation, carried out on the basis of dialogue with the participants through individual and group interviews. During the expansion of the subjective vision of the effects of meditation on emotional intelligence hermeneutical approach takes center stage, implying that personal experience is hidden and revealed through reflection, which appears to stimulate participants' interview with the researcher (Ponterotto, 2005). Thus, the qualitative approach to data collection is preferable for deeper understand the personal experience and its estimation by respondents.

Thematic and interpretive phenomenological analysis are the key methods of the obtained data interpretation within the epistemological stance of the researcher stated above. The main qualitative method in this research is the interview. Semi-structured interview with the primary use of open questions was chosen as the primary means of data collection for subsequent analysis. It promoted free and comfortable thinking of the respondents, and description of the individual experience in their own terms (Smith et al., 2009). The balance of key themes of primary and secondary interviews differed due to objective conditions of the research and the necessity of wide discussion of questions on the meditative experience and dynamics of the emotional intelligence owing to meditation practice at the secondary interview. Each participant was interviewed by the researcher twice.

The technique of the emotional intelligence estimation by D. V. Lyusin - "EmIn Questionnaire", based on the author's concept of the emotional intelligence has been used during the research for the quantitative measurement of the emotional intelligence and its dynamics in the participants of both experimental and control groups in the beginning and in the end of the experiment. D. V. Lyusin defines emotional as an ability to understand own and others' emotions and their management (Lyusin, 2009). The emotional intelligence as an ability to understand and to manage emotions according to Lyusin can be directed both at own and other people's emotions. Thus, the author actually differentiates concepts of intrapersonal and interpersonal emotional intelligence by actualization of different cognitive processes and skills connected with each other (Sergienko and Vetrova, 2009).

Research Participants

The experimental part of the research took place in October - November, 2013 and involved 48 Russian-speaking migrants from the former Soviet Union at the age of 19 to 54 years, residing in the territory of modern Germany being at the initial stage of migration (up to 5 years of stay in the host country). The choice of this group of migrants as a target one is caused on the one hand by specific issues of the social-psychological adaptation in the group, and on the other - typical problems of migrants' adaptation. It should also be noted that the limitation of the current research focus of this group of migrants was caused in order to achieve greater depth of data analysis (Smith, 1996), which also corresponds to the selected research methods. In addition, being in the initial stage of migration, contributing to the highest load on the psyche of migrants and their level of emotional intelligence, will take into account the most acute psychological problems. Thus, sample was made by the migrants in basic life changes of a social environment and experiencing significant pressing on the mentality due to objective life circumstances. Participants of the study were divided into two groups: experimental, consisting of 32 people and includes two subgroups practicing 20 minute transcendental meditation and mindfulness meditation daily and a control consisting of 16 people. Thus, all participants in the study were actually divided into 3 subgroups of 16 people in each (9 women and 7 men) that represent age and sexual characteristics of the target group in general. The control group representatives did not learn meditation (confirmed by the poll) and did not practice any kind of meditation that was periodically checked during the research. Their emotional intelligence was measured by the chosen technique similarly with the participants of the experimental group. Thus, the representative sample was made by randomization of the participants into experimental and control groups to minimize possible restrictions, able to result in the experiment irrelevancy.

Results

Interpretation of the interviews and "EmIn Questionnaire" data to assess the primary level of emotional intelligence. The analysis of the primary interview data allowed to reveal that motivation is determining the degree of mental health indicators, emotional comfort, the level of development of emotional intelligence key competencies, the choice of individual coping strategies, and ultimately the efficiency of the social and psychological migrants' adaptation in general. Most of the participants have no meditation experience, which should contribute to obtaining adequate and valid assessment of the impact of meditation on emotional intelligence as a key factor in the social-psychological adaptation. The participants have significant issues of social-psychological adaptation and effectiveness of emotional intelligence competencies, ranging due to the subjective and objective factors. The existing complexity of adaptation due to the presence of psychological problems and manifests in the emotional sphere of migrants. Meditation could be potentially effective tool for influencing the emotional intelligence as a key factor in the social-psychological adaptation of migrants.

Both quantitative and qualitative methods showed higher levels of intrapersonal component of emotional intelligence in study participants with simultaneous low levels of interpersonal component. According to the researcher, it may be associated with both objective and subjective factors. It is also due to the fact that the established interpersonal communication is destroyed during the migration process. In connection with that a large part of the internal forces of man rushes to introspection, self-reflection, which leads to a greater awareness of their own feelings and emotions, what participants talked about in the interviews. Also it should be said that a significant relationship changes with simultaneous reduction of community, isolation from the indigenous population and some difficulties in dealing with other migrants, largely mediate the problem of adequate formation and positive dynamics of interpersonal component of emotional intelligence. The followed interpretation of the data led to the conclusion that the baseline of emotional intelligence on the scale of interpersonal and intrapersonal components in the experimental and control groups are similar quantities and the difference in quantitative terms is within the statistical error, which allows to use these results as a baseline for the objectives of this research and to disseminate the results to the target group as a whole.

Interpretation of the secondary interviews and "EmIn Questionnaire" application. Interviews with participants of the experimental group, practicing mindfulness meditation and transcendental meditation, have shown some differences in subjective meditation experience. For example, respondents who practiced transcendental meditation in their responses to interview questions noted primarily the following changes: calming the thoughts during meditation; stopping of the internal dialogue in the process of meditation; perception of the meditation sessions as a rest from the stream of life and from their own problems; awareness of the meaning of life, the emergence of insight in the process of meditation; normalization of physiological symptoms of stress and anxiety (heart rate, palpitations, sweating etc.). Respondents who practiced mindfulness meditation also noted certain changes observed in the process of meditation (stopping the internal dialogue, the sudden decision of problems during the session (awareness of a possible withdrawal from the situation), the perception of the meditation as a recreation, etc.), but primarily they noted the following changes: a greater awareness in everyday functioning; normalization of relations with relatives and friends due to a greater understanding of their needs; general "degree reduction" of emotions in relationships with others and everyday issues. Thus, we can conclude that the experimental group participants practiced transcendental meditation individually evaluated the internal changes as a priority, while respondents who practiced mindfulness meditation largely focused on external positive changes as a result of meditation practices. In general, most of the participants of the experimental group praised their meditative experience.

Most of the experimental group participants reported some improvement characteristics of interpersonal competencies of emotional intelligence. Significant differences between the answers of respondents who practiced different types of meditation, as well as significant discrepancies in the subjective assessment of the effects of meditation on the interpersonal component of emotional intelligence have not been traced. Thus, we can note the definite improvement in emotional intelligence interpersonal component that was subjectively associates by the respondents with the practice of meditation. Simultaneously the relative stability in intrapersonal competence of the emotional intelligence with a slight tendency to strengthen was demonstrated with significant positive dynamics in control emotions and expression control due to the practice of meditation. In fact, most respondents noted a significant strengthening of effective coping strategies as the basis of social and psychological adjustment. Overall assessment of the social-psychological adaptation prospects became more positive as compared with the primary interview data that the majority of respondents associated with positive changes in their own emotional sphere due to the practice of meditation. It should be also said that the analysis of the secondary interviews with the participants of the experimental group showed a positive trend of emotional intelligence levels in the context of respondents' interpersonal and intrapersonal EI components both in subjective evaluation of participants and according to the researcher's opinion. In addition, respondents definitely associated the positive changes with the practice of meditation compared to its state at the time of the primary interview. It's important that the choice of a specific type of meditation showed no significant correlation with the benefits of the meditation practice.

For objective assessment and mitigation manifestations of subjective factors the analysis of secondary interviews with participants in the control group which did not practice any kind of meditation throughout the experiment was performed. Control group participants did not show any significant changes in the subjective assessment of the EI levels, both in interpersonal and intrapersonal components. There were also no any significant changes in assessing the prospects of adaptation, the general prospects in life and mood in the control group participants observed as compared with the primary interview.

As a result of the analysis of the obtained "EmIn Questionnaire" values of key components of the emotional intelligence at the research termination, the following conclusions have been made. The experimental group participants showed definite growth of values of the interpersonal component of the emotional intelligence comparing to the primary measurement. Growth of these values is shown with a simultaneous drop of the share corresponding to a low level of the interpersonal emotional intelligence. Comparison of initial results of measurement of the intrapersonal component of the emotional intelligence with results of the secondary measurement in the experimental group allows to judge about their relative stability with a simultaneous tendency to positive dynamics of the intrapersonal emotional intelligence due to meditation. Additionally for comparison of influence of the transcendental and mindfulness meditations, the data have been generated by two parts of the experimental group participants, which practiced the specified kinds of meditation. Comparison of the primary and secondary measurements of levels of the key components of the emotional intelligence by meditation types, practiced by the experimental group participants allowed to make a conclusion that the choice of the certain meditation types does not affect significantly on the positive dynamics of the emotional intelligence values. Dynamics of values of the intrapersonal component of the emotional intelligence for different meditation types also is comparable. Like general sample results, by both meditation types, a relative stability with a tendency to increased average indexes of the intrapersonal component of the emotional intelligence was observed. For the purpose of comparison experimental and control groups EI values at the end of the research components of the emotional intelligence were measured in the control group.

Dynamics of the key components of emotional intelligence in the control group was compared with the dynamics, traced in the experimental group. It need to pay attention to the relative stability of the primary and secondary EI interpersonal component levels in the control group with simultaneous slight tendency toward redistribution of low and medium shares. Comparison of measurements of the intrapersonal component of the emotional intelligence in the control group also testifies the absence of any significant changes. Thus, it should be noted absence of significant changes and dynamics of key values of the emotional intelligence in the control group during the research that confirms a hypothesis about factorial impact of meditation on the emotional intelligence as the key factor of the migrants' successful social-psychological adaptation. The obtained results of interpretation of the qualitative and quantitative techniques mutually confirm the obtained data that testifies the result validity and reliability also as research internal consistency.

Conclusions

1. Analysis and interpretation of the data obtained in the practical part of the study revealed that meditation affects the dynamics of emotional intelligence. Findings from the application of qualitative (interviews) and quantitative ("EmIn Questionnaire") techniques clearly demonstrate the positive dynamics in the experimental group.
2. Dynamics of the emotional intelligence that arises due to the practice of meditation is positive. Results of the analysis of the primary and secondary measurements

of interpersonal and intrapersonal components of the emotional intelligence in the experimental group participants and interpretation of the interviews showed significant positive changes of these parameters comparing to the data in the control group. The respondents have definitely indicated that meditation helps to bring a sense of order and stability, improves the ability to concentrate and focus on reality simultaneously suppressing negative thought forms that allows to raise the overall positive attitude.

3. Meditation practice strengthens key intrapersonal and interpersonal competencies of the emotional intelligence, causing efficiency of the migrants' social-psychological adaptation. Considerable positive dynamics of the interpersonal competence of the participants' emotional intelligence has shown in the empirical part of the research at a simultaneous positive tendency in dynamics of the intrapersonal component of the emotional intelligence in the experimental group participants. The analysis of the interview with the respondents has shown certain subjective correlation between the emotional intelligence and efficiency of the migrants' social-psychological adaptation. Most participants of the experimental group noticed that the meditative practice helps to discipline the mind and to become more conscious in demonstration and expression of feelings and emotions. Positive meditation influence is expressed in decrease in the stress, adequate estimation of own forces and objective restrictions that mediates the effective social-psychological adaptation.

4. The choice of a specific method of meditation does not affect essentially on the dynamics of the emotional intelligence. The conclusion is confirmed by measurements in the experimental group regarding the practiced meditation type. The mindfulness and transcendental meditations were offered to the participants in the research. The analysis of the interview with the participants showed some differences in estimation of the individual meditation experience (high concentration on internal factors in the transcendental meditation practitioners and emphasis on external signs of mindfulness meditation impact). This trend has been shown at similar values of the dynamics of emotional intelligence in both parts of the experimental group and data interpretation of the subjective assessment of respondents' meditative experience and its impact on emotional intelligence in the context of social and psychological adaptation.

Thus, the research confirmed that meditation is an effective tool of positive influence on the migrants' emotional intelligence, mediating success and efficiency of their social-psychological adaptation, which expresses an integrated scientific novelty of this work. The obtained data about a positive influence of meditation on the migrants' emotional intelligence is especially important under globalization and growing number of cross-cultural contacts. Besides, the research results allow to consider the meditation as an effective method of influence on the emotional intelligence, including for the practical counseling psychology and psychotherapy at work with the migrants, one of groups of the modern society, experiencing the greatest psychological stress. Meditation techniques introduction in the counseling psychological practice, of course, requires the further in-depth studies of meditation as a factor of impact on emotional intelligence, but at the same time the data of the present study permit to begin practical approbation of meditation in psychotherapy of the migrants.

The obtained empirical data were confirmed by an adequate theoretic-methodological base of the research, the author used valid and reliable techniques of the qualitative and quantitative estimation of the research subject area, selected by direct conformity with a theoretical and methodological substantiation, the author's epistemological position and reflected aim, objectives and formulated hypotheses. We considered restrictions in previous researchers and tried to integrate the quantitative and qualitative techniques that allowed to raise the research relevance and reliability of the received results.

Implications of Study

The practical value of the research is presented by possibility of application of its results for the aims and objectives of the counseling psychology, including at work with migrants as the positive impact of meditation on the emotional intelligence in a context of the migrants' social-psychological adaptation is empirically confirmed during the research. The author believes that the results of the research can promote the further studying of a role of meditation in development of the emotional intelligence as meditation can be used as one of the most effective tools focused upon increase of meditators' adaptable possibilities.

Because the concept of the emotional intelligence is a new scientific category, the further scientific researches of factorial influence on the emotional intelligence components are required in addition to specification of key concepts, in a context of the migrants' social-psychological adaptation, including studying of meditation impact on the migrants' emotional intelligence. The more detailed studying of meditation is required to solve this problem, as well as formation of the uniform scientific concept of the emotional intelligence by the structure and key components, development of an effective technique of the emotional intelligence quantitative measurement, able to provide valid and reliable data for the factorial influence analysis, and further deep studying of connection of the emotional intelligence w with a number of personal characteristics and indicators of success in modern society.

References

1. Andreeva, G. M., (2004). Social Psychology. Moscow: Aspect Press.
2. Bennett, J. M.; Bennett, M. J. (2004), "Developing intercultural sensitivity: An integrative approach to global and domestic diversity", in Landis, D.; Bennett, J.; Bennett, M. J. (eds.), Handbook of Intercultural Training, Sage Publications, Inc.
3. Braun, V., & Clarke V.. (2006). Using thematic analysis in psychology. Qualitative Research in Psychology. 3 (2), 77-101.
4. Busygina, N.P., (2009). Phenomenological and hermeneutical approaches to qualitative psychological researches. *Cultural-historical psychology*. 1, 57-65.
5. Chiesa, A., (2010). Vipassana meditation: systematic review of current evidence. TheJournal of Alternative and Complementary Medicine. 16 (1), 37-46.
6. Howitt, D., (2010). Introduction to qualitative methods in psychology. London: Pearson.
7. Goleman, D., (2001). The Art of Meditation. Audiobook 2CD. New York: Macmillan Audio.
8. Lyusin, D.V., (2009). EmIn Questionnaire: new psychometric data. Moscow: Institute of Psychology of Russian Academy of Sciences.
9. Pankova, T.A., (2011). The role of emotional intelligence in the socio-psychological adaptation of young professionals. *Psychological Research*. 4 (18).
10. Patel, N., (1999). Getting the Evidence: Guidelines for Ethical Mental Health Research Involving Issues of "Race", Ethnicity and Culture. London: Mind Publications.
11. Ponterotto, J. G., (2005). Qualitative research in counseling psychology: A primer on research paradigms and philosophy of science. Journal of Counseling Psychology. 52(2), 126-136.
12. Ryan, G. W., & Bernard, H. R., (2000). Data management and analysis methods. In Braun, V., & Clarke, V. (2006). Using thematic analysis in psychology. Qualitative Research in Psychology. 3, 77-101.
13. Sergienko, E.A., Vetrova, I.I., (2009). Emotional intelligence: Russian-language adaptation of The Mayer–Salovey–Caruso Emotional Intelligence Test, Version 2.0. *Psychological Research*. 6 (8).
14. Smith, J., (1996). Beyond the divide between cognition and discourse: Using Interpretive Phenomenological Analysis in Health Psychology. Psychology & Health. 11, 261-271.
15. Smith, J. A., Flowers, P., & Larkin, M., (2009). Interpretative Phenomenological Analysis: Theory, Method and Research. London: Sage.
16. Zherebtsov, M.V., (2004). Method «Grounded theory» as a method of qualitative data analysis. *The Moscow University Herald*. 18 (1), 89-90.

Music as an Effective Method of Teaching English as a Foreign Language at School

Maria V. Arkhipova
State University of Nizhniy Novgorod, Russia

Abstract. *The article is devoted to the research of the problem of increasing pupils' motivation while learning a foreign language at school on the basis of music. It presents the analysis of possibilities of using music aiming to optimize an educational process. It introduces an original program of raising teenagers' motivation to learn the English language by means of music.*

Keywords: *motivation, school age, foreign language, musical art.*

1. The problem of learning English as a foreign language at school and means of music as a way to motivate it

The questions of raising motivation and interest to the foreign languages' learning are one of the most acute problems in modern pedagogics and psychology. This topical matter is connected with the social and cultural situation nowadays when the foreign language is becoming an important factor of economic, scientific, technical and cultural progress of society. That is why the problem of raising motivation and interest to the foreign languages' learning is considered very significant.

Foreign languages have great opportunities to educate students in a complex way. Foreign languages in modern society are characterized as a means of interpersonal, intergovernmental and international communication. That is why a question of motivating pupils to study English is of vital importance nowadays.

The problem of motivation appears from the first steps of studying at school. Though there are a lot of works devoted to this problem (V.G.Aseev, V.S.Ilin, A.K.Markova, J.W.Atkinson, S.P.Grossman, K.B.Madsen, G.Merphy and others), it cannot be characterized as being worked out. The question of finding effective methods, being able to heighten interest to a foreign language, is very significant today.

The researchers of motivation problem point out to its decrease from class to class (I.N.Andreeva, 2002; I.A.Zimnyaa ,1991; I.B.Minaeva, 2009; and others). The researches show the fact that before the moment of learning a foreign language and at the very beginning of it the students at school have a high level of motivation. They want to speak a language, recite poetry and sing foreign songs. But with the beginning of studying the language the attitude of the students to it changes. A lot of pupils become disappointed with it because this process supposes to have a large period of learning a basic material, a stage of primitive contents, the overcoming of different difficulties. All these peculiarities move aside the achievement of aims which were dreamt by the pupils at first. A low interest of studying a language remains up to the moment, when pupils begin to understand the necessity of knowing this subject for the future education and profession.

Except mentioned features, the physiological characteristics of the pupils are also considered to be the reason of a low level of motivation. Among these characteristics are the fear of making a mistake and thus to lose prestige among the classmates, fatiguability with the routine form of the lessons. So some psychophysiological traits, which are necessary to take into account during the lessons of a foreign language, can be defined:

- a great fall of motivation of studying a language after the first year of learning it;
- aversion to authoritarian forms of work;
- fear to make a mistake;
- boredom to monotonous forms of lessons.

These facts show that it is necessary to find such methods of teaching English which would best suit physiological and psychological needs of the pupils and which would form a wish of learning a language. Research suggest different strategies of doing it, such as the creation of the situation of success (Oxford J., 1994), the organization of collective teaching (Dornyei Z., 1994), the use of bright material (Salanovich N.A., Gorbyshina O.V. and others), the organization of team-work (Perevozchikova V.N. and others) etc.

But now the idea of using music at the lessons of the English language attracts the attention of teachers, as it is music that influences the emotional sphere of pupils, and emotions are known to be motives of any activity.

A lot of scholars pay attention to music, and they have an undivided opinion that music is an art that has the greatest power of emotional impact. As it is noted by N.V.Shutova, the impact of music is felt by everybody. The impact may be stimulative and sedative. All these factors made people include music as an obligatory subject to the system of education long ago [1].

Facts proving a stimulating influence of music have been attracting scientists' attention for many years. Interesting observation and experiments have been made in this field. Thus it is a well known fact that in Ancient China some illnesses were treated by means of music. In Ancient Greece, Rome and Egypt the art of music achieved a high level. Pythagoras believed that music made people healthy. In works of Plato the idea that music harmonizes a person with society was expressed. Aristotle introduced a conception of purification of soul in process of art perception.

In the XX century the music therapy began to be widely used in Europe as a cure of neuropsychic and somatic illnesses. Among specially organized centers the following ones became the most popular: in Austria – "Osterreichische Gesellschaft zur Forderung der Musiktherapie", Switzerland – "Schweizer Forum fur Musiktherapie", Germany - "Arbeitgemeinschaft fur Musiktherapie".

At the beginning of the XX century the experiments showed that perception of music fastens heartbeats, increases rushes of blood to brain, slows or fastens pulse. There are also many researches proving that music intensifies metabolism, effects muscular tonus, stimulates the appearance of emotions. Music at that time began to be used during the educational process. L.N.Tregubova points to the fact that music used at lessons makes pupils more attentive. The investigations of V.I.Petrushin show that instrumental music stimulates imagination [7].

Among modern works, devoting to the problem of using music in educational process, it is necessary to mention the researches made by N.I.Shutova. The author presents facts illustrating the ability of musical impact to produce a positive emotional state. The following information has been proved by the scientist:

- purposeful use of music stimulates the appearance of emotions, improves mood of a person;
- heightens efficiency of work during the lessons;
- increases memory capacity;
- activates creativity;
- increases the desire to study [1].

The idea of using music exactly at the lessons of a foreign language attracts attention of the scientists and teachers. At present time the musical material can be found practically in all school-books. But the authors more often propose to use music during the out-of-school activities, and the question of using music as a basis during educational process is not widely covered in literature. Different aspects of this problem we can find in some works of the scholars. V.F.Aitov supposes authentic songs to be an integral part of teaching English. Especially it is more effective in the primary school, because in such a way children begin to learn and understand the culture of a studied country. Moreover songs stimulate creative thinking and form a good taste. From the methodical approach a song is viewed as an example of a foreign speech and as cultural information [5]. Music is widely used by O.V.Kudravetz practically during the whole lesson:

1) for a phonetic exercise at the beginning of a lesson;
2) for learning lexical and grammatical material;
3) as a stimulus for the development of speech habit and skills;
4) as a kind of relaxation in the middle or at the end of a lesson, when pupils are tired;
5) for making pupils less tense;
6) for recreating a capacity to work [4].

Thus, it should be noted that a wide range of researches is devoted to the problem of using music in teaching English. But the works, describing music as a basic element in teaching English (where music is not an additional material but a main one) are not numerous. In most cases the use of music is reduced to listening and singing the alphabet, songs and what's more is that the emphasis is made on the primary school.

2. Results of the pilot program of motivating pupils to learn English on the basis of music

First we made a research of pupils' motivation level learning English as a foreign language by traditional methods. It showed that only 18 % of pupils have a high motivation to learn English. The majority are characterized by a low interest to the subject, neutral emotions to the process and result of studying. 36% of pupils suffer from emotional discomfort at the lessons, feel not confident and ill at ease.

The results of research prove that most pupils have low motivation to learn the language. Thus we elaborated on the program aiming to make pupils feel eager to learn English.

The aim of the pilot program was to work out and run an experiment raising motivation of English learning at school by means of music. Our experiment lasted one year, the components of which being described in the table below.

TABLE 1. Components of units aimed to raise motivation of learning English by pupils

Component	Aim	Lesson stage	Music techniques
Emotional and controlling	Reduce emotional discomfort, develop positive emotional attitude to a lesson	Introduction - creating atmosphere of a lesson	1) vocal greeting 2) music relaxation 3) music psycho-gymnastics
Intellectual and motivational	Raise productivity of a lesson, success of activities	Main part - realizing lesson's objectives	1) passive music perception 2) music dramatization 3) music performance
Resulting and evaluative	Low the fear of failure, assure confidence of success, develop motives of positive activity results	Conclusion - comprehension and result consolidation	1) passive music perception 2) music performance

Motivation of learning English had positive changes in the result of our pilot program. Control study showed that the number of pupils having a high level of motivation increased from 18% to 62%. Positive emotions and interest to the subject had a considerable growth.

We can conclude that music used in the process of teaching English as a foreign language at school has a positive impact on the process of studying. Interest and positive emotions created by music have a good influence on the results of pupils' activity. Motivated pupils strive for success by getting thorough knowledge of English, taking the initiative and thus acquiring linguistic independence that leads to desired effects.

The purposeful use of music can produce positive emotions that would optimize the educational process. Positive emotions contribute to the increase of memory capacity, good knowledge of teaching material, the raising of creative activity, thus the motivation of learning foreign languages increases and the quality of teaching improves. Consequently it is possible to motivate, stimulate and optimize the foreign languages' learning by using music.

References:

1. Arkhipova M.V., Shutova N.V. Music as an effective means of raising motivation to learn foreign languages. - LAP LAMBERT Academic Publishing, Saarbrucken, Deutschland, 2012
2. Zimnaya I.A. Psychology of teaching foreign languages at school. – Moscow, 1991
3. Iljin E.P. Motivation and motives. – Saint Petersburg, 2006
4. Kydravetz O.V. Music and songs at the German language lessons // Foreign languages at school. – 2001. - № 2. – p.45-50
5. Minaeva I.B. Motivation of success as a necessary condition to know foreign languages // Foreign languages at school. – 2009. - № 2. – p.42-44
6. Nikitenko Z.N., Aitov V.F., Aitova V.M. Authentic songs as one of the culture elements to teach a foreign language in primary school // Foreign languages at school. – 1996. - № 4. – p.14-21
7. Petrushin V.I. Musical psychotherapy. – Moscow, 1997
8. Grossman S.P. The biology of motivation. – Annual Review of Psychology, 1979, Vol.30, p.209-242
9. Moore D.G., Burland K., Davidson J.W. The Social Context of Music Success: A Developmental Account. – British Journal of Psychology, Vol.94, 2003, p.529-550

Genius Sleeps inside Everyone, or What Is Real Genius: Pathology or Destiny and How to Become a Genius

Anna Dukhareva
Alexey Mayorov

OOO "VectorPeople", Moscow, Russia

Abstract. *This article is about the latest research in psychology and neuropsychology on issues of talents and their higher degree of development — genius. The authors have analyzed the psychological development of talented persons, and compared it with the psychological development of normal people. You`ll read about the organization of our psychological life and ways of checking for deviations in the different stages. The main part of the article is dedicated to the issue of self-growth and the disclosure of personal abilities and talents.*

Keywords: *talent, genius, pathology of development, self growth.*

What is the secret of genius? Why do some people create masterpieces, which touch the depth of the minds and souls of their contemporaries and descendants, whereas others can't reach such heights? For many years people have studied biographies of Einstein, Beethoven, Aivazovsky and other great persons in order to understand their secret. The results of these studies are scientific and popular-science manuscripts about the nature of talent and genius.

Now we are experiencing the latest turn of the popularity of genius. In the age of mobile and Internet technology we imitate genius in different ways. People repost quotes of famous people on the social media web, actively visit training sessions, and read books about self-development. The popularity of genius has not spared child psychology. Bookstores' shelves are full of books which explain how to make your child a genius, and a lot of private pre-schools promise to teach your son or daughter to speak three languages before he's one year old.

But does the imitation of genius get us closer to it? That doesn't really seem to be the case. And the abundance of information that we get in training sessions? We don't really hear about geniuses who were produced through training or coaching sessions. Besides, if we teach skills to a child when his brain structure is not ready, we won't be making a genius out of him, but will be holding up his development in the next stages.

So what makes a genius great? Why does someone, for example Mozart or Pushkin, create a lot of wonderful masterpieces, while others, for example, Dante Alighieri, are famous for only one of their achievements? And what must we do to be close to the top? We propose to look at this and a lot of others questions on genius from psychological and neuropsychological viewpoints.

What is genius? We suggest that readers learn some principles of work on the psyche to find the answer to this question.

The psyche is a subjective reflection of the objective world. The psyche is a specific function of the brain. Current neuropsychology doesn't consider questions of dichotomy, "mind-body" or "soul-body". C. G. Jung said that deep inside the unconscious there is a point where psyche and body are one. Any process of the psyche has a reflection in the work of the nervous system. Geniuses are no exception.

The psyche has the following structure (Table 1):

Unit of neuropsychological analysis of human behavior is a factor. This concept was offer the "father" of neuropsychology A. R. Luria. Factor is a reflection of specific aspect, a link in the holistic psychology system, function and process. Condition of a factor can tell us about normal and pathology in a human behavior. For example, deficit of the reciprocal coordination (coordination of difficult movements, associated with antagonistic working muscles) is the pathology factor of hemispheric interaction in the kinetic praxis.

The psyche has a long time of development, which has its beginnings in the mother's womb and doesn't end. But all the structures of the brain structures which are responsible for psychological work are formed by the age of 21.

There are a lot of mistakes which can appear in a child's development. Psychological development is a difficult process, which is influenced by:

- child`s heredity
- health of parents
- course of pregnancy and childbirth
- social situation of development (who educates the child, whether or not he/she has siblings, the style of family upbringing, who looks after the child, etc.)
- etc.

There are all manner of failures that can appear in this difficult process. They can be caused by an unfavorable combination of genes, a difficulty pregnancy, birth trauma, childhood communicable illnesses, incorrect educational measures.

Moreover, problems of typology and variability deviations complicate the individual features of each personality. That's why borders between norm, upper and lower borders of norm, and, of course, pathology are very difficult to find.

But what is genius if not an upper level of pathology? Of course, at first this statement is deters, and we want to argue with it. We offer to make it clear in a definition of pathology.

TABLE 1. Structure of the Psyche

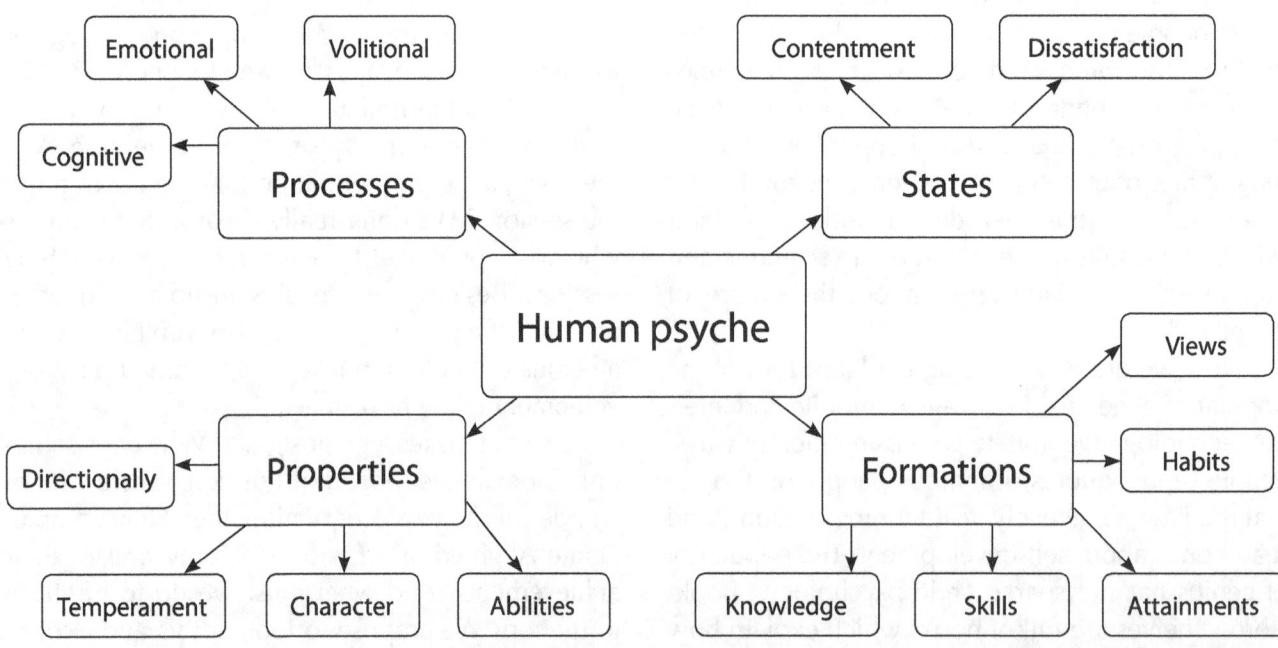

"Strictly speaking, there is nothing in pathology (super-norm) which would not have been in norm and vice versa. <...> Norm has the three basic parameters of differences from abnormal variety (in any sides) of self-actualization:

- quantity, quality and severity of pathology — or super-norm phenomena per unit time;
- their actual syndromic unity evident in the pathology and vice versa, mosaicism, lack of systematization in norm;
- the ability of arbitrary self-regulation, including the invariant registry funds attracted from the outside and finding an «ecological niche». That means, finding the mechanisms of adaptation to his behavior by a person (relating to the mainstreaming of this phenomena) and environmental requirements" [1, 94–95].

That means that people who belong to the super-norm or upper pathology demonstrate this in their behavior; often compared with the norm they display some factor, systematic of their display, with an ability to control this factor.

There are a lot of people who have musical talent. But musical genius is different with its special sensitivity for the height of musical sound and harmony and its combinations. This ability combines with a love of music, systematic education and self-education and the ability to control one's own musical hearing (there would not have been a lot of benefit from Rakhmaninov if he had not been able to compare harmonic sounds at his own behest).

It seems as if we have lifted the veil hiding the secret of genius. This is the super-norm, with one or more factors responsible for a specific activity. We more or less know the reasons of pathology, but how can we develop the supernorm?

The supernorm is a super adaptive specific factor or group of factors, which takes responsibility for one activity. Many people would argue with this approach, for example: "Hey! We know that all geeks are cranks, what kind of adaptability may be involved?!" But super-adaptive may only be relevant in some factors, while all others working are pathological.

Let`s recall the theory of J. Piaget on intellectual development, in particular the concepts "circuit" and "pattern". Circuits are unique intellectual acts, which can adapt for the purpose of changing the environment. Patterns are repeating actions. Let's give a simple example.

We usually get up, turn off an alarm clock and go to a bathroom. We get used to this ritual. We know where our toothbrush is, from what tap hot or cold water flows, where our towel is. Getting washed in the morning doesn't need circuits. Moreover, it would be very uncomfortable and difficult to develop a new way to go to the bathroom, to brush our teeth or turn on the shower every day.

But now imagine — horror of horrors!!! — the hot water has been shut off. You won`t be ready for this and discover the problem when you turn on the hot water tap. But the absence of hot water won't change the fact that you must wash (actually, will you go to work with unbrushed teeth?). Now you need a circuit: how to wash without hot water and not get ill, because not everyone can take a very cold shower in the morning.

Imagine that a genius is a person who thinks only in circuits in specific ways of working. We can say that he is superadaptive in this kind of working, nothing can knock him out of the track. Meanwhile, in others kinds of working, which seem to us ordinary, among geniuses all may be forced by patterns. We know stories about these guys, the weirdo-geniuses. Let's recall how the great physicist L. D. Landau nearly died because he forgot to eat. But we must point out that freakish behavior isn't a mandatory property of genius. There are a lot of examples of this: A. Einstein, V. A. Mozart, L. S. Vygotsky, etc.

Thus, a genius is a person with specific brain structures, developed to an upper level super-norm, that let's them be super-adaptive in specific kinds of working.

It seems that we could shed light on the mystery of genius. But the reader's curiosity won't be fully satisfied until we touch upon the question of the development of genius.

Let`s return to Table 1: Structures of the Psyche. In total every structure of the psyche determines how we see reality and how we interact with it. The psyche of the genius has the same structures as the psyche of every another person. But if we are saying that genius is an unusual activity in a specific kind of working that means that we must learn more about the properties of the psyche.

Psyche properties (PP) are sustainable educations, which provide specific levels of human working. This is exactly those structures of the psyche which directly influence what, why and how we do something. PPs include: direction, temperament, character and abilities. Also PP can be shared between three categories depending on those participating in psyche energy exchange: properties of knowledge, properties of communication and interaction, properties-guideline.

To the energy of the psyche we refer ideas, information, existential meanings and emotional resource. Properties of knowledge are PPs which influence cognitive working, i.e. its properties which are responsible for mining the energy of the psyche.

Properties of communication and interaction are the person`s qualitative influences on activities for harnessing energy of the psyche, i.e. its implementation in an individual's productive activities.

Properties-guideline are those same filters which monitor the quality of energy which we use.

Except for supernormal development, some certain factors which make the genius superadaptive in some kinds of working, they have developed PPs which provide them with a large volume of psychological energy that makes a genius very productive.

Genius is not an idiosyncrasy, but a phenomenon caused by abnormal development of a human`s psyche. Genius includes:

- factor of certain working or kinds of it, that development refers to as the upper border of normal;
- developed groups of PPs, which let them work at a very high level in certain kind of work and…
- … ability to operate the large volume of psychological energy.

Want to become a genius — vain, perhaps, but not bad. But imitation of the great doesn`t help. You can smoke a pipe and bring yourself to RFK light and consume opium and get put in a mental hospital but you won't become Sherlock Holmes. If you are photographed with a protruding tongue you won`t become Einstein, either. Current training sessions don't make you a genius: they have other goals and development programs. And don`t try to make a little genius of your child by being guided by the cheap benefits offered by child psychology.

We want to finish our article with some thoughts about how to become a genius. Genius is the realizing of potential, which is pledged inside everyone. All of us have an ability to become a great person. "…speaker, there is nothing in pathology (supernorm) which would not have been in norm and vice versa" [1, 94]. You must concentrate on looking for a job, in which you really want to excel, it may be nuclear physics or motherhood. You`ll not develop separate skills, like they do with coaching and training, but psychological properties. A part of those are congenital, like temperament or abilities. Proceeding from this you can choose your job. But direction and character are the properties which may be developed. A strong character can cover weaknesses of temperament, and directionally can push you to work which you weren't inclined to do at the beginning.

It`s important to understand that personal development is often out of the usual scope of being, like geniuses are out of commonness. "The genius is uncomfortable in his nature. But people like comfort", said Robert Walser, Swiss poet and writer. Therefore when you start the journey to your hidden qualities, ask yourself: "Am I ready to go out from my comfort zone, and how far am I ready to go?"

References

1. Semenovich A. V., "Introduction to the Neuropsychology of Childhood", 2013.
2. Rubinstein L. S., "Fundamentals of General Psychology", 2002.
3. Iacoboni Marco, "Reflected in Humans", 2011.

New Approach for Integration of Psychological Knowledge

Anton Karasevich
Belarusian State University, Minsk, Belarus

Abstract. *We first created Four-Dimensional (Space-Time) Model of Psyche. This model integrates all existing psychological paradigms. On the basis of the Model is possible to solve a wide range theoretical and application problems. This article represents the foundations of our new concept.*

Keywords: *Four-Dimensional Psychophysical Model of Reality, Dinamic Model of Psyche, modelling of psyche.*

Four-Dimensional Psychophysical Model of Reality is a promising new approach for the integration and systematization of worldwide psychological knowledge that can be successfully used in general, social, medical fields of psychology. Using the Model we can integrate the basic psychological paradigms to unite their practical potential, and to find new areas of perspective research. Psychology has accumulated huge amount of scientific knowledge and skills. The problem is this knowledge is not systematized. We can do that with the use of Four-Dimensional Psychophysical Model of Reality. Visualization makes knowledge understandable and easily accessible. The computer program, which is now developed on the basis of the Model, can be used to model all diversity of psychic processes.

The Model enables joint description of psyche, human organism and environment (Figure 1).

"Desire-goal" matrix is the basis of Dynamic Model of Psyche. It includes the time period of mental activity between activator of mental activity ("desire") and "brake" of mental activity ("goal"). "Desire" and "goal" in The Model are identical with many different notions of psychological paradigms.

Under our Model there are external (organismal) and internal (psychical) activities. Organismal activity includes the organism movement and work of organs. Psychical activity includes the processes of sensation, perception, internal speech, thought, imagination, memory and emotion. It corresponds

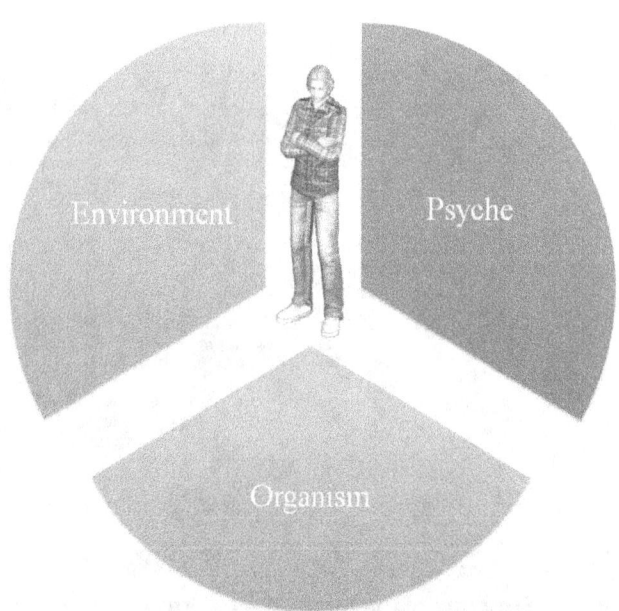

FIG. 1. Components of the Model

to the dynamic block of psyche. Stationary block of psyche covers all inborn and accumulated psychical material of the human. At the instant only small part of the psyche is in the psychical processes (actualized part). Other biggest not actualized part of the psyche is a stationary block of psyche. Stationary block has many components: knowledge, needs, goals, motives, temperament, character, self-esteem, ways to respond, and many others.

The Model consists of three elements: dynamical and stationary blocks of the psyche, organism (organismal activity) (Figure 2). The one is three-dimensional and dynamic. It is graphically made with *3ds Max*. The Model can demonstrate all psychological paradigms and social processes.

Four-Dimensional Psychophysical Model can be applied for modeling of:

- **human psyche, brain and behavior** (Figure 3);
- **psychological theories.** For example, Polikarpov's theory (Theory of Temporal Feedback) that explains the possibility of the future prediction (Figure 4) [1];
- **physical theories.** For example, Fontana's theory (The Four Space-times Model of Reality) and Shulman's theory (Spherical Expanding Universe Theory) that explain structure of the Universe (Figure 5) [2, 3].

We briefly have modeled only three point of time in the context of human activity, but for a detailed modeling we need to analyze the huge number of such moments. Necessary to create a computer program to demonstrate the dynamics of the whole of human activity. This program will take into account the full range of determinants of mental activity. This program will be able to describe in detail the structure of the psyche, human physiology, his behavior and the environment in general. The creation of this program is a priority direction in the development of the Model, because the simulation of single sections is time-consuming and complicated work. The integration paradigms must be accurate, compact, laconic, otherwise the integration will not give anything except a methodological confusion. Our Model is simple, but it includes everything needed for modeling a general psychological scientific knowledge. The Model especially designed maximum open. For example, the number of components of a stationary block of psyche and their contents are not strictly defined. This enables to find a lot of areas of intersection of scientific knowledge. A detailed version of the Model (with a demonstration of the dynamics of the nervous system) integrates psychological knowledge and neurobiology. Also, in the integration of scientific knowledge, we can describe diversity of its philosophical foundation. It reinforces the general methodological foundation of Model.

Let us describe the Four-Dimensional Psychophysical Model of Reality using the figure (Figure

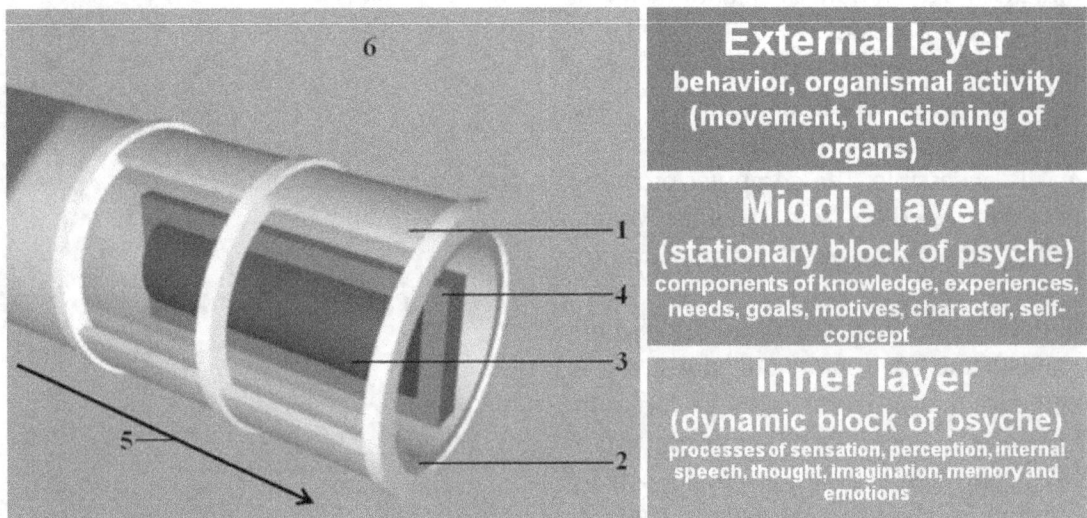

FIG. 2. Structure of the Model: 1 – external layer (denotes behavior and organismal activity), 2 – "desire-goal" matrix, 3 – dynamic block of psyche, 4 – stationary block of psyche, 5 – growth direction of the Model, 6 – environment

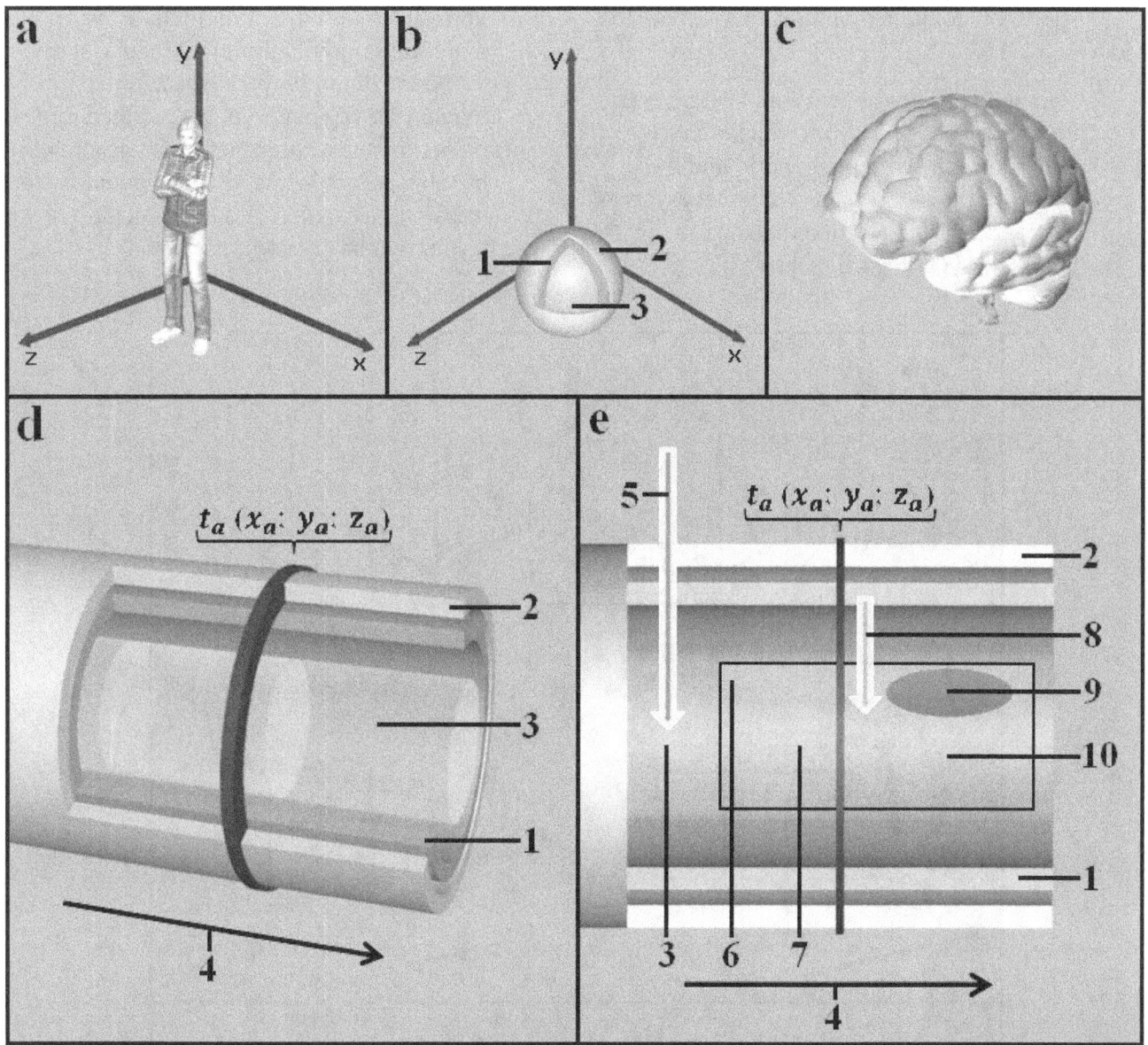

FIG. 3. Structure of Four-Dimensional Psychophysical Model of Reality: 1 - stationary block of psyche, 2 - external layer of model (denotes behavior and organismal activity), 3 - dynamic block of psyche, 4 - growth direction of model, 5 - transition of the external world into the psyche, 6 - process of thinking, 7 - other psychological processes involved in map of consciousness, 8 - influence of stationary block of psyche on psychical activity, 9 - conscious part of the dynamic block of psyche, 10 - unconscious part of the dynamic block of psyche

3). Let us take modeling of single point in time in the context of the overall activity of human. This is an example of a possible structure of the Model. In the figure indicated: a – location of human in the space at the moment t_a (this figure may include an human's environment), b – sphere model (shows content of the human psyche at the moment t_a), c – model of human brain (we can describe physiological processes), d – general model of psyche, e – dinamical model of psyche, side-view, with the map of consciousness (shown only a process of thinking).

Perspectives of the Model application:

- Creating a computer program based on the Model;
- Integration and systematization of global psychological knowledge;
- Interdisciplinary dialogue with the use of the Model.

Versions of computer program based on the Model:

- **"Home Psychologist" version.** Will be used for personal use. Everyone will be able to understand the own inner world and the causes of the all own actions, to get psychological counseling via Internet, and to create the own psychological map;
- **Version for business.** Will include testing of employees and monitor their mental state for optimization of working capacity;
- **Version for science.** Will be intended for professional psychologists. This version will be a big psychological guide. Conditions for research, discussion and interpretation of results will be created.

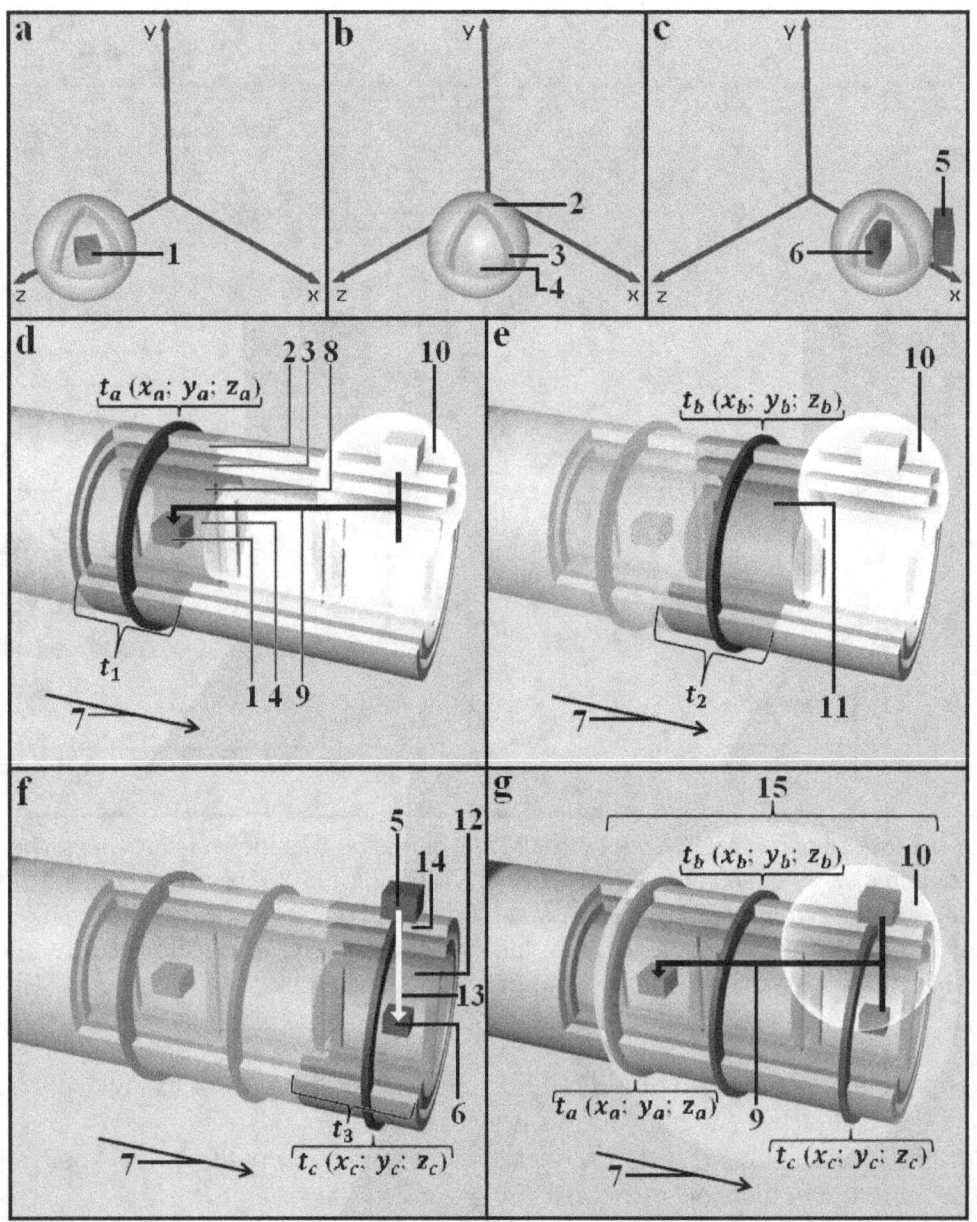

FIG. 4. Polikarpov's Theory of Temporal Feedback: 1 - index of an event, 2 - external layer of the model (denotes behavior and organismal activity), 3 - stationary block of psyche, 4 - dynamic block of psyche, 5 – an object related to the future event, 6 - an object into the human psyche, 7 - growth direction of the model, 8 - the first "desire-goal" matrix (sleep), 9 - temporal feedback, 10 - attractor, 11 - the second "desire-goal" matrix (thinking of sleep after waking up), 12 - the third "desire-goal" matrix (an event related to the object), 13 - transition of an object to the human psyche, 14 - organismal activity associated with the object, 15 - "space" of an event

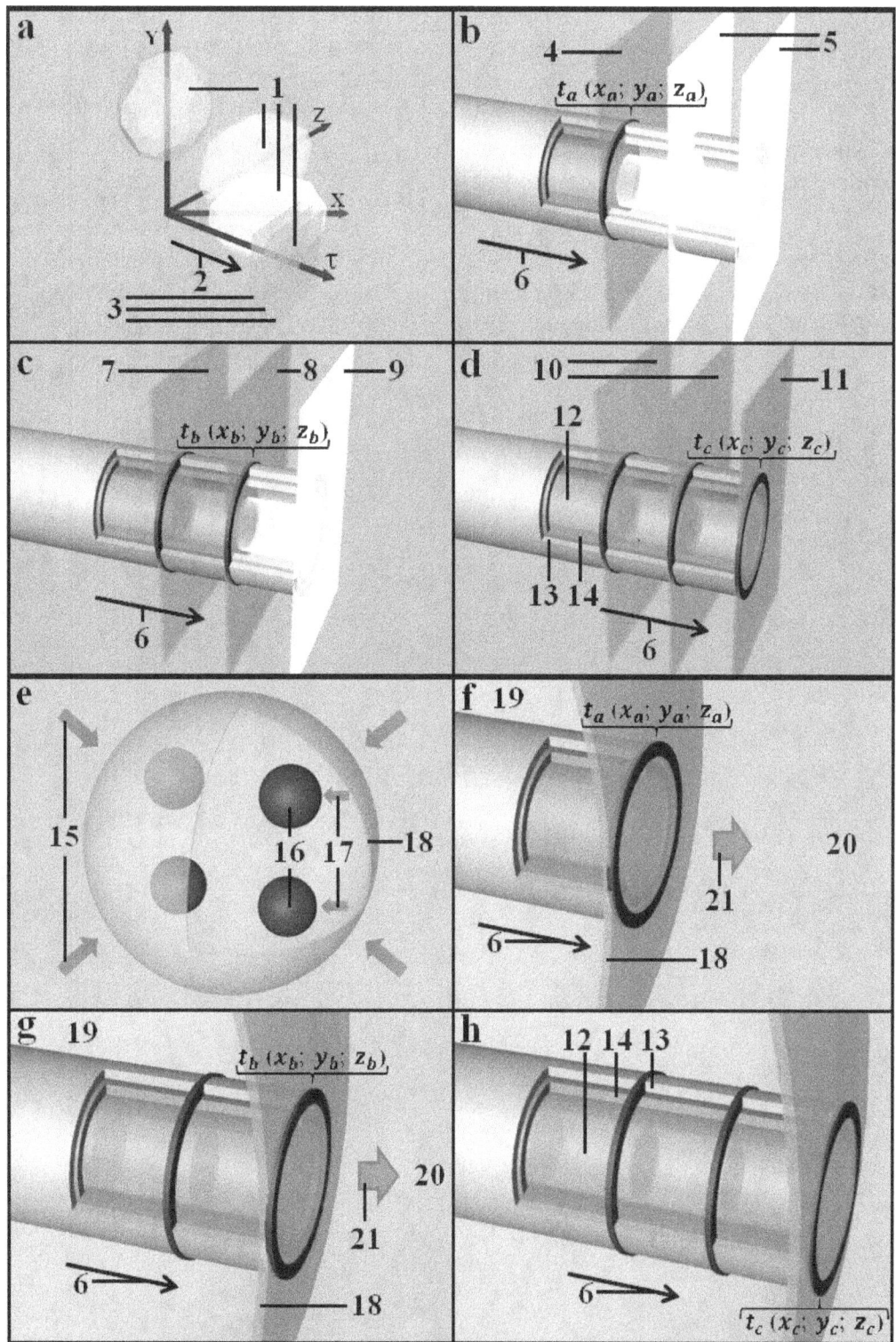

FIG. 5. Fontana's theory and Shulman's theory: 1 - four space-times, 2 - motion our universe along τ-axis, 3 - continuum of τ-frames, 4 - τ-frame is the present, 5 - τ-frames are the future, 6 - growth direction of the model, 7 - τ-frame is the past, 8 - τ-frame is the present, 9 - τ-frame is the future, 10 - τ-frames are the past, 11 - τ-frame is the present, 12 - dynamic block of psyche, 13 - external layer of the model (denotes behavior and organismal activity), 14 - stationary block of psyche, 15 - inflow of matter, energy and information from white hole towards black hole, 16 - black holes, 17 - influx of energy to the internal supermassive black holes, 18 - our universe is a three-dimensional membrane (three-dimensional black hole) between outer a four-dimensional super-universe and a giant black hole, 19 - space of a giant black hole, 20 - super-universe (white hole), 21 - direction of sphere increasing

Four-Dimensional Psychophysical Model of Reality may cooperate with physics, biology (in particular neurobiology), sociology, philosophy, and other sciences. The Model can demonstrate extensive scientific knowledge and explain human behavior. The Model can create new practical theory.

References

1. *Polikarpov V.A.* Theory of Temporal Feedback (in Russ.) / Library of Electronic Publications, Institute for Time Nature Explorations, Lomonosov Moscow State University. http://www.chronos.msu.ru/RREPORTS/Polikarpov_teoria-temporalnoi.pdf
2. *Fontana G.* The Four Space-times Model of Reality // AIP Conference Proceedings. 2005. Vol. 746 (1). P.1403.
3. *Shulman M.H.* Cosmology and metabolism (in Russ.) / Library of Electronic Publications, Institute for Time Nature Explorations, Lomonosov Moscow State University. http://www.chronos.msu.ru/RREPORTS/shulman_metabolizm.pdf

The Problems of Formation of Social-Psychological Adaptation and Communicative Competence in Students of Higher Education Institutions

Guli S. Salomova
Bukhara State University, Bukhara, Uzbekistan

It is not a secret that a human character is erratic and diverse and no one doubts that he has individual qualities and specific "styles". But it does not mean that the diversity of a human character is limitless, because humans' mutual communication, interrelation, their merging into various social groups is a condition of preserving warm psychological environment.

Goziev E.G states that " Each individual during his youth is a master of specific social relations and is the object of many economical, political, legal and behaviorial influences" [1].

The period of teenage years which is full of spiritual feelings and mark the whole life of a human is one of the main scientific objects of many psychologists and pedagogicians. During teenage years the main features of a human being form, and it is observed that on the basis of it values, traditions, and other social aspects emerge.

One of the most complicated psychological processes of teenage years is an effective course of adaptation process and a formation of communicative competence.

The system of accompaniment of social- psychological adaptation of the young people towards education in higher education institutions successfully works in all developed countries, contributing to harmonious entrance of young specialists into the professional environment. An effective model of psychological-pedagogical support of the young people in multiethnic environment is a main tool in solving the problemsput forward in the Conception of educationmodernization in all countries [2].

The critical point of young people's transition from one educational environment to the other must be accompanied by a qualified assistance, capable of easing professional adaptation of them to the social-economic situation, characterized by the gap between "demand" and "offer", between the degree of professional competence of young specialists and the demand of employers [3].

We consider **socio-psychological adaptation** as the process of an individual's coming into a new social environment and his familiarization with it. Differing from biological adaptation,in socio-psychological adaptation there is a unity of adaptive and transformable activities. Psychological traits and behavior of a man transform in the response to the demands of social environment and in the course of meeting demands of adaptation the social environment changes itself [4].

In higher education institutions the students come across with several problems. In case these problems are not positively solved by teachers and other responsible people they will cause such unpleasant situations as poor acquisition and the division into subgroups within the group.

The duration of adaptation period determines their present and future success. The adaptation of students while studying in higher education institutions and during extra curricular activities is a phenomenon related to the radical change of a person's social state. The complexation of the activity takes place and young men and girls enter the system of relations (traditions, steriotypes, values and others) which is new for them [5]. Moreover, it changes their understanding about their future life and the features of a new social micro environment.

Taking into consideration that this radical continental psychological change during the student years plays a negative role in the formation of spiritually perfect person the psychological service of the institution, group supervisors and teachers must work towards increasing the students' adaptation and conduct a research on this problem.

Here it is important to diagnose and assist them on time. There are many methods and experiments for the diagnostics of adaptation.

First of all we can use the scale of K. Rogers and R. Diamond to determine socio-psychological adaptation. Other methods are designed to accompany it. For example: the methodology to determine aspiration and self-evaluation. (T. V. Dembo, S.L Rubenstein, A. M. Prikhojan); questionnaire with 16 aspects (R.Kettel); Bass-Darky questionnaire, social competence questionnaire (L. M. Mitina); motive constructivenessquestionnaire (A. A. Rean, O. P. Eliseeva)(Rean); the questionnaire which determines self attitude (V. V. Stolin, S. R. Pantileev); " Control localization " (adapted by E. G. Ksenofonova); the diagnostics of communicative social competence) (N. P. Fetiskin, V. V. Kozlov, G. M. Manuylov); "The socio-psychological features of the communication subject" (V. A. Labunskaya) [6].

"The express diagnostics of personal traits" worked out by Kettel, Eisenk, Lichko and others has also proved helpful. It diagnoses the position of the students in interpersonal relationships (extravert, neurotic) and provides with the instructions for teachers on how to deal with the identified types of students.

The methodology "The study of the students' attitude to subjects" worked out by Kazanseva for students proved very effective.

This situation often results in the errors in distribution of study and free time, the process of professional adaptation slows down and the feeling of self-dissatisfaction emerges. The difficulties of interpersonal character appear. As a result the necessity to solve the matters of differentiation of conflictual knowledge in the frame of polyethnic environment of the young people and to realize the peculiarities of conflict emergence at the early stage of adaptation of studying in higher education institutions comes forth.

Having considered the actuality of the problem, the matters of formation of conflictual competence in the frame of a linguistic institution, which develops the communicative competence of polyethnic environment of the youth are of interest [7].

In the process of interrelation between students in the polyethnic environment of the institution conflicting situations occur from time to time, which can be solved positively or grow into a bigger conflict. At an early stage of education at a higher education institution each student has his own idea about goals and means of their realization, plans and principles oforganizing student life. That's why it is natural that the prime element of professional competence of the contemporary student is his conflictual competence, which includes in itself the collection of specific knowledge and skills.

Thus, we can define the main components of a freshman's adaptation process as following:

a) the acquisition of new study norms, grades, the fulfillment ofindependent works and other requirements, learning the types of scientific activity;
b) the adaptation to the new group and its traditions;
c) the adjustment to the dormitory, to the new conditions of a lifestyle, to the new patterns of student life, to spending free time in a different way.
d) the formation of a new attitude towards work

There are many types of social adaptation. They are: disadaptation andsuch psychological processes as active and passive.

Disadaptation is characterized by the vagueness of objectives and types of the human activity, it is defined by the rejection the norms and values of the new social environment, and sometimes working against it [8].

In *passive adaptation*, the student accepts principles and norms under the slogan "I am like everybody else" and does not try to change anything, though he is capable of it. Passive adaptation is seen in the presense of simple objectives and easy types of activities. But the range of relations and the scale of problems that have to be solved in this case differ from desadaptation degree [9].

Active adaptation, first of all is capable of fulfilling complete socialization successfully. The student acquires not only the norms and traditions of the new environment, but also starts to see newer types of various goalsduring this process. To realize these goals the student searches tirelessly, establishing the active system of relations with the people who are associated with his professional future [5]. Khowing his own place in society, goals and obligation he will not deviate from his way. At last this degree of adaptation will form a proportional unity with himself, the world and the people.

Besides this, nowadays there is a problem of increasing the students' communicative competence by teachers and group supervisors. Moreover, it differs from the formation of adaptation and is considered to be the phenomenon which doesn't depend on students at some degree.

According to A. P. Prosetskiy, adaptation is a process and a result of the internal change of the external active skills under the new condition of an individual's existence [8].

The issue of formation of student's communicative competence and the methodology recommended by the experimental psychologists will be introduced at the next stages of the work.

According to the researchers the reasons of the freshmen's having difficulty in acquisition is not their poor preparation at high schools and colleges. The reason of it is their personal traits, the personal readiness to study, the ability to gain knowledge independently, self control and assessment, the ability to plan their time accordingly have not yet formed fully [4].

Many students come across with some difficulties related to the lack of independent work experience during the first lectures or practical lessons. It will be obvious that they cannot make notes during lectures, cannot work independently with textbooks, cannot gain extra knowledge from the primary sources, cannot analyze a big deal of information and finally cannot express their thoughts clearly and distinctly [10].

Everybody knows that the education at a higher education institution totally differs from education at a high school and other educational institutions. The student must always prepare for the classes at high school. If he does not prepare he will surely get "a two". From the day he begins attending university classes he will listen to the lectures first and then practical classes begin. The students think that they may attend the practical lessons without a preparation. Thus there is no obligation to read, to learn by heart, to retell. Just no one requires it. As a result, the first semestr seems very easy, and the students develop an understanding that it is possible to study the material of the past lessons just before exams and the student approaches the lessons indifferently.

But when he fails exams, and cannot pass daily tests and midterm, the student who is not industrious and has no high motivation looses self-confidence and interest in studies [7].

Teenage age is the stage of self realization, formation of personal outlook, making important decisions that determine the person's later way of life. In an ideal situation a teenager strives for choosing a profession that denotes his place in society and the meaning of his life. Together with the formation of the feeling of self-realization, there emerges a desire to come into contact with the environment on the basis of trust. This determines the quality.

The need of teenagers to communicate with adults doesn't always end positively and it interferes with the complete development of a teenager's personality at some extent. Here the role of a group supervisor is very important. This situation requires attentiveness on behalf of a supervisor. He provides a healthy communication of students with their surroundings. He must also be able to imagine to rely on the human resource emanating from personal opportunities.

In the formation of a student community, it is important to take into consideration the process of group dynamics peculiar to each study group. Differing from the adaptation process the group supervisor plays an important role in the formation of a communicative competence. The process of adaptation is a phenomenon that depends on the student. The healthy communicative environment in the group and the formation of the competence depend on the teacher and supervisor. We can observe the formation periods of the group:

1. The period of orientation(adaptation). This period is a time of group formation and organization. In this period each individual tries to find his place in the group, is engaged in finding the answers to the questions like "How should I introduce myself?", "Will I be accepted in this group?", "Whom will I like?". As the students don't know each other well, they are afraid to feel active and attacking. As a result, the degree of a conflict in the group will not be high.
2. The period of a conflict(activation, high activeness). In this period, each student puts his interests above others and tries to find the answers to the questions "Who will influence on me from this or that aspect?", "How can I satisfy my interests and needs?". The period ofresponsibility distribution runs and during this time each student puts forward his own point of view. The students openly go into conflict with each other. This process is an emotionally agitated process and the emergence of several subgroups within the group is observed. At this period it is possible to control the conflict without ending it or keeping far from it.
3. The period of unanimity formation (constructive, active movement). After the conflicts the group directs its energy to solve the problems in the group and collect specific norms of behavior. The above mentioned misunderstandings and obscurities emerge in the group again. But these consist of selecting measures to prevent conflicts.
4. The period of consensus or support. The group functions as a whole working group. Reflection, making decisions will be seen on the basis of the group unanimity. But it does not consciously eliminate negative emotions in order to form the constructive rival feelings in itself. If the group does not reach unanimity, it may stop at the first or second stage.

Conducting meetings devoted to the end of the first term and academic year. This event is conducted as a meeting with students and their parents. Moreover, conducting the psychodynamic activities with students is also considered. Within the frame of a psychodynamic approach game therapy and art therapy are organized. Behavioral approach is connected with the formation of optimal models of adaptive behavior and behavioral skills through teaching, developing compliant competence(training, behavior), modification of conduct through changing the way of thinking (cognitive therapy), developing self regulation(skill-therapy).

Communicative training as an active process of academic interrelation of students and teachers is not only delicate and respectful form of communication, but also has some specific content of effective adaptation to the study in an institution. It allows to get over many crises of age development and effectively forms a personality of a student.

References

1. Tulaganova, G. K. Difficult teenagers. T., 2005
2. Kunitsina V.N.,Kazarinova N.V., Pogol'sha V.M. Interpersonal communication. SPb.:2001
3. Kryajeva, I.K. Social-psychological factors of adaptation of the worker in industry.//Practical problems of social psychology. M., 1983.
4. Rean, A.A., Kudashev, A.R.,Baranova.A.A. The psychology of the adaptation of a person. SPb., 2002.
5. Khanchuk N.N. Some actual problems of students' adaptation in the course of studying at higher school// The problems of social adaptation of various groups of population under modern conditions.-Vladivostok: Far the Eastern university publication, 2000.

6. Under the publication of A.A.Rean, The psychology of a teenager.M., 2007.
7. Boronina, L.N., Vishnevskiy Yu.R., Didkovskaya Y.V., Mineeva S.I. The adaptation of the first year students:problems and tendencies// University administration. 2001.№4(19). P65-69
8. Prosetskiy P.A. Psychological peculiarities of students' adaptation to the conditions of study at the higher education institution . M:2009.
9. Krisko, V.G. Social psychology: Dictionary-reference book. (The library of practical psychology).-Mn.;M.:2001.
10. Kuznetsova, N.V. The conditions of the adaptation of the first-year students to the academic process in Blagoveshensk branch of SGA [Electronic resourse]/ N.V. Kuznetsova.-The regime of the access:www.muh.ru.

About Relationship Between Defense Mechanisms Of The Mentality and The Level Of School Anxiety Of Senior Pupils

Natalia Turan

Kemerovo State University, Novokuznetsk, Russia

Abstract. *This article presents the results of the research of interrelation of defense mechanisms and the level of school anxiety of senior pupils.*

Keywords: *defense mechanisms, school anxiety, interrelation of defense mechanisms and the level of school anxiety.*

The constant changing of the word acts as a long stressful factor, it difficult for many people to adapt; this difficulty can lead to an increase in anxiety. Such a specific kind of alarm as school anxiety may appear in adolescence. An emergence of school anxiety associated with the socio–psychological factors is the most typical. A receptivity to these factors increases during the adolescent crisis of senior pupils.

The term "defense mechanisms" in psychology denotes unconscious mental processes aimed at minimizing the negative experiences. The protective mechanism is in the processes of resistance.

A definition appeared in 1894 in the work of S. Freud "The Neuro-Psychoses of Defense" and was used in his future works for the description of the battle between "I" and painful or unbearable thoughts and affects. Later, this term was abandoned and replaced by the term "superseding". However, the relationship between the two concepts to this day is unclear [1].

In Russian psychology, conceptions of defense mechanisms are also studied. One of the conceptual approaches to psychological defense is the approach of F. V. Bassin. According to his point of view, the psychological defense is the most important form of reaction consciousness of a person`s psychological trauma. Another approach is in the works of B. D. Karsavarskij, who considers psychological protection a system of adaptive reactions of personality, aimed to protect the change of significance of a non-adaptable relation`s components (cognitive, behavioral, and emotional) and to weaken their traumatic impact on the "I-concept" [4].

Nowadays in Russian and foreign scientific literature, there are discussions about what kind of defense mechanisms are considered less primitive and immature, how to distinguish them, and how many of them there are. For example, B. V. Zeigarnik distinguishes the destructive and constructive measures of protection. Nancy McWilliams allocates two levels of defense mechanisms, primary and secondary, which include such protection as denial, suppression, regression, compensation, projection, substitution, and intellectualization.

The problem of anxiety from the psychological point of view was put out for the first time and was subjected to review in the works of S. Freud. In this case, his views on the matter of anxiety and fear were close to the views of the philosopher I. Kierkegaard. They both recognized the need to distinguish between fear and anxiety, considering that fear is a reaction to concrete danger, while anxiety is a reaction to unknown and undetermined danger. S. Freud defined anxiety as an unpleasant experience, presenting a signal of future danger. The result of anxiety is a feeling of uncertainty and helplessness [1].

School anxiety is a rather mild form of expression of the emotional problems of a child. It can be expressed in agitation, in increased anxiety in school routine, in forms, in an unpleasant attitude to themselves, or in negative appraisal of teachers and other pupils. The pupil feels inadequacy and defectiveness; he is not sure about the correctness of his conduct or his decisions [2]. A. M. Prikhozhan believes that the emergence and consolidation of the anxiety associated with the dissatisfaction of the leading age learner needs which acquire excessive nature. The anxiety becomes a stable personality formation in adolescents with the help of the features of the "I-concept" and relationships to themselves. Before that, it was a derivative of a wide circle of family violations [3].

The issue of the interrelation of school anxiety and defense mechanisms has been studied with the help of the test questionnaire about defenses mechanisms "Life Style Index" (LSI) (R.Plutchik, H. Kellerman, H. R. Conte; (Adaptation E. Romanova, L. R. Grebennikov)), a test questionnaire of dominate defenses mechanisms in communication (V.Boyko), the scale of reactive and personal anxiety (Spielberger-Huning), and the test for diagnosing the level of school anxiety (Phillips).

According to the results of the defense mechanisms analysis (test questionnaire LSI), dominate defense mechanisms are regression and repression. Such defenses as projection, reaction formation, and compensation have a medium level of severity. Such defenses as denial, displacement, and intellectualization have a low level of severity. This means that pupils don't deny the information or facts which they couldn't bear and understand. Also, they didn't unconsciously substitute acts, thoughts, and needs to more safe and rational things.

The study of dominate defense mechanisms in communication (the V.Boyko test questionnaire) have shown that the dominate defense mechanism in communication is an aggressive strategy as a demonstration of instinctive defense behavior to save personal reality. The communicative strategy of avoidance has a medium level of severity, meaning that probationers prefer to escape the stress and conflict situations but don't defend their own points of view. The peacefulness psychological strategy stays on the lower level of severity, because the character and personality of senior pupils is not formed yet. This is a normal phase of personal development, because adolescence is the most critical period of personal development, which results in the integral personality being formed.

The diagnostic of the level of school anxiety was studied with the help of the scale of reactive and personal anxiety (Spielberger-Huning) and the test of the level of school anxiety (Phillips). According to the results of the study of reactive and personal anxiety by scale (Spielberger-Huning), senior pupils have a medium level of reactive anxiety, meaning that respondents have the ability to assess the situation and to make a decision rationally; the decisions are not only influenced by emotion.

Also, we diagnosed the level of school anxiety of senior pupils with the test for diagnostic of level of school anxiety (Phillips). The diagnostic showed that probationers have a medium level of school anxiety: pupils have the ability to perceive school situations adequately, stimulating their own activity and the activity in the right direction. Most of the respondents had a medium level of fear of inconsistencies in their expectations of social surroundings. In other words, pupils tend to focus on the significance of other people in the assessment of their own actions and thoughts, causing an anxiety of estimations, expectation of negative comments and opinions from others, and an acute reaction to constructive criticism. Also, we have documented a low level of the test scale "Problems and fears in a relationship with teachers". This indicates a favorable emotional background in relationships with adults at school, which allows pupils to learn successfully.

We documented the Pearson correlation analysis ($\alpha = 0{,}413$, $p \leq 0{,}05$) of the relationship of school anxiety and defense mechanisms.

A significant correlation relationship presents interrelation between the scales: "displacement" (test LSI) and "aggression communicational strategy" (test V. Boyko) ($r = 0{,}45$, $p \leq 0{,}05$). Analysis presents significant a negative correlation relationship between the scales: "Low physiological resistance to stress" and "Problems and fears in a relationship with teachers" ($r = -0{,}42$, $p \leq 0{,}05$) (Philips' test of school anxiety). Also, we found that the scale "Reactive anxiety" (test Spielberger-Huning) has significant correlation relationships with the scale "Frustration needs to achieve" (Philips' test of school anxiety) ($r = 0{,}44$, $p \leq 0{,}05$) the scale "Social stress" ($r = 0{,}44$, $p \leq 0{,}05$) (Philips' test of school anxiety) and the scale "Social stress" has significant correlation relationships with the scale "General school anxiety" ($r = 0{,}47$, $p \leq 0{,}05$) (Philips' test of school anxiety).

The results of the research of the interrelation of defense mechanisms and the level of school anxiety of senior pupils will be interesting and useful to psychologists at professional practices, because knowledge of interrelation of defense mechanisms and the level of school anxiety of clients helps the specialist to optimize client behavioral strategies. Also, the issue of the interrelation of school anxiety and defense mechanisms we studied will help teachers and parents to create a favorable atmosphere at school and in the family, because psychological health is the most important aspect of personal growth and development.

References

1. Freud, A. (1937). The Ego and the Mechanisms of Defense, London: Hogarth Press and Institute of Psycho-Analysis. (Revised edition: 1966 (US), 1968 (UK))
2. Methodical journal «School Psychologist» [External link]: http://psy.1september.ru/
3. Prichozhan, A. M. (2000). Children's and teenager's anxiety: psychological nature and dynamic, Moscow: MPSI, NGO «MODEK», p.304. ISBN: 5–89502–089–5, 5–89395–174–3
4. Romanova, E. S., Grebennikov, L. R.(1996). Defense Mechanisms: genesis, functioning, diagnostics, Mytischi: «Talent» Publishing, p.144

SOCIOLOGY

Impact of the Demographic Structure of the Population on the Stability of an Urban Family

Fania Igebaeva
Bashkir State Agrarian University, Ufa, Russia

The city's population representing its specific subsystem operates and develops in the system of economic and social relations, the nature of which is determined by all the socio- economic structure of society. However, the population of the city is linked by a common residence, common social and material resources of its life. Like any social and territorial community of people, the city is constantly undergoing not only economic and social but also demographic changes. These changes constitute the contents of the demographic development of the city – the transition process from one qualitative state of population to another.

Demographic development of a society including cities covers a wide scope of human activity and is not limited by purely demographic facts and processes of fertility, mortality, number of marriages, migration, changes in sex composition etc. The social character of the demographic development is manifested in the fact that it is carried out in the course of changing the social qualities of the population – the level of education, health, capacity limits. However, the renewal and changing of social relations and relations between groups, generations (in particular intra-family relations) can be traced.

One of the reproduction features of the urban population is its relatively unstable demographic structure. If in rural areas the proportion of younger people and older people has remained stable over the years, in urban areas, depending on their size, location, age, economic growth, variations in terms of the demographic development are considerable.

High dynamic of socio-demographic structure of the city sometimes leads to distortions in the sex ratio and age groups. This affects the level of marriage and fertility in its turn, as far as many young people can not marry because of lack of marriage partners. Of course, sooner or later, married couples will be formed, but the optimal age gap between brides and grooms will be broken. Studies show that a harmonious relationship in the family depends on the ratio of the age of husband and wife. According to some family researchers the age gap between spouses should be 4-6 years [1].

The optimal age gap between spouses is conditioned by both biological and social reasons. Women live longer, but get older earlier, social maturity of men comes later (years of study, service in the army postpones the time of entry into independent life. Therefore, if the women early marriage can be somehow reasonable, the same can not be said about early marriages of men. A situation where the husband is much older than his wife or wife is older than her husband in most cases leads to sexual disharmony, and as a consequence, to disharmony in family life in general. Disharmony in the marital relationship is generated by both social factors and physiological features of the spouses' development. Dissatisfaction with sexual life in marriage is more common due to subjective reasons; however, it is obvious that sexuality is one of the main values of marriage.

The predominance of women in the population of the city leads to one more very important moral consequence: a significant number of men entering the second or subsequent marriage to young women who have not been married. For example, according to statistics in Bashkortostan second and subsequent marriages were concluded by 12-14 % of men and 9-10 % of women among all new marriages. [2] If we consider that men getting married for the second time in many cases (approximately 35-40 %) take younger women as their partners, but divorced women rarely get married to men who have not been married, it becomes evident that gender disparity turns into a kind of men's "polygamy". This is facilitated by the fact that a growing number of divorced men and women do not come more in officially registered marriages, though they live as an unmarried couple. This is confirmed by the census data, when there are much more married women than married men. The discrepancy is explained by the fact that living together as an unmarried couple women more often consider themselves married than their partners. In addition, it should be noted that men become widowers half as often (annually about 300 thousand of men become widowers in the country) as women, and not more than one third of men, whose marriage broke up last year, remarry once more annually. It is also known that in recent years there is a growing number of men who are not married (either legal or in fact).

It should be noted that the violation of the proportions in the demographic structure of the city not only leads to a reduction in the marriage and fertility levels, but also gives rise to such specific social phenomenon as "rivalry" of girls and women in the creation of marriage unions. If we add to that part of the women, who are doomed to "unmarried life" due to lack of marriage partners, and even women who are divorced and not re-married (in Ufa there are about 7-8 thousand of such people), then forms a significant contingent of female population that falls out of the process of reproduction of new generations. In addition the excess of unmarried women has a psychological impact on fragile families, creates additional conditions for infidelity, and reduces the level of claims to potential marriage partners (men). In particular, the ease with which divorces occur in cities can be explained not only by the simplified procedure of divorce but also by a psychological confidence in a choice of another marriage partner. So, in big cities the probability of divorced men to remarry is three times higher than among women. [3] "Rivalry" among women as a result of smaller opportunities to get married to a certain extent stimulates the increasing of extramarital affairs, and as a result we get non-marital births.

"Distortion" of demographic structure of the population of the city, and therefore fluctuations in the mode of reproduction of the population is essentially conditioned by the nature of migration flows, rushing in and out of the city. The impact of migration mobility of population on family stability is mainly mediated. The increased percentage of youth migration flows not only leads to disturbances in the proportions of the sexes, but it is itself a destabilizing factor for the existing families by increasing the anonymous connections, degree of sexual partners choice etc.

Professional, social and spatial mobility of citizens is objectively necessary because without it, new areas are impossible to be developed. But at the same time migration processes affect fertility. Selected studies made by USATU sociological laboratory, show that there are differences in the family sizes between the indigenous and non-indigenous citizens. Worse living conditions among migrants, delayed "demand" for children, and moments of adaptation and restratification certainly lead to economic and socio-psychological difficulties of migrant families. Thus, the strengthening of social and professional mobility affects not only fertility, but also the stability of the urban family. In our opinion, dialectically controversial nature of the relationship of migration mobility of the population and family stability should be noted.

The reverse effect of marriage and family relations state on the demographic development of the city primarily can be traced in the fact that the degree of harmony depends on the level of family relations and marriage migration mobility of the population. We mention that the migration is caused by the forced nature and dissonances in the marital relationship which not only enhance the territorial mobility of the population, but also adversely affect the timing of marriage, the birth of children.

The very fact of moving to another city makes many families correct their plans. Unhappy family life and divorces can act as additional incentives for moving home. Migrants polls held in Ufa certify that the destabilization of family relationships enhances migration mobility of the population of the city. Approximately 20% of potential and 30% of real migrants call "family circumstances" the defining motivation to move - moving to children (parents), marriage, and the "escape from the family". Thus, to the question in our questionnaire "Do your family move with you?" only 33.7 % of respondents said yes, 10% said that the family will not move at all (and 50% of them directly pointed to the "family troubles"). [4]

Thus, we can conclude the following: the destabilization of the family somehow leads to delayed demand for children, the desire to change the place of residence, and this in its turn leads to the spread of nuclear families, single mothers and fathers, as well as violation of the optimum ratio of the age between brides and grooms.

Family instability directly affects the level of fertility, as far as tensions in the marital relationship; lack of confidence in the marriage partner is often decisive in the question how many children the family will have. To a greater extent, it does not refer to the first and the second and especially the third child whose appearance can be seen as an answer to the question of how things work in the family. The evidence of this, even being indirect, is the fact that according to our study materials of registry offices in Ufa (900 couples) only 5.6 % among divorced had three children, 52% had one child, and 12% do not have children.

Children in the family contribute to its consolidation, since they increase the responsibility of spouses between each other and the society. At the same time the more developed the educational function of the family is, the more stable and harmonious it develops. The nature and status of marriage and family relations, the level of family stability depends on its value. This was particularly evidenced by the results of our survey. Thus, in families where marital relationships are more stable and harmonious, there tend to be more children than in families with fragile and strained marriage and family relationships. Lack of harmony in the marital relationship, marriage dissatisfaction with extremely high divorce rate is inversely proportional to the level of fertility. Of course, one can not ignore the impact of general social and economic factors and demographic structure of the population on fertility. But we cannot ignore the degree of stability of marriage. As noted by L.E. Darski the "increased probability of marriage dissolutions led to a more low-level marital fertility as a woman is afraid of being alone with children and spouses do not want to "bind" themselves by a large number of children taking into account the potential possibility for divorce"[5]. Therefore, according to many experts on marriage and family one of the most important conditions for increasing the birth rate in addition to stimulating measures of demographic policy in this country (talking about maternity capital) is to prevent divorces, family separations, increasing the stability of marriage and family relations [6].

References

1. Igebaeva F.A. Features of demographic development of the city and family stability. // Social and Political Sciences. - 2013. – No. 2.
2. Demographic processes in the Republic of Bashkortostan. Statistics compilation. 2012.
3. Antonov A.I., Medkov V.M., Archangelskyi V.N. Demographic processes in Russia in the XXI Century, Moscow, ID "Graal", 2002.
4. Igebaeva F.A. Impact of migration on the reproductive behavior of citizens. // Science, education and society: problems and prospects of development. International scientific and practical conference materials. - Tambov, 2013.
5. Darski A.E. Demographic statistical study. M., "Statistics", 1999.
6. Igebaeva F.A Sociology. Manual. - Moscow: INFRA-M, 2014. – p. 236

The Specific Features of Uzbek People's Ethnoculture

Maksuda S. Khajieva
Mangubek Urazmetov
Sayyora Akmanova

Urgench State University named after Al-Khwarizmi, Urgench, Uzbekistan

Ethnic culture (ethnoculture) includes the material, spiritual, daily, and educational wealth of all people and nations. Ethnoculture is created in the process of historical development and also understanding the world, assimilating and changing experiences. It is difficult to supply national culture development without learning about the wealth and experiences and without using positive sides in them.

"The value of history and historical epoch doesn't mean it depends on a long past but also it means to help for solving modern problems. If we learn the history of origin of modern problems and find their foundations we'll find the means of solving them"[1]. So, historical insight to the problem is not only necessary for reconstructing historical processes but also for realizing the essence of modern problems and solving them effectively. Scientific-theoretical conceptions, which are directed at learning about the Uzbek people's ethnoculture, can be divided into the following categories: historical-archaeological approach, mythological-religious approach, social lingual approach, economical approach, folklore approach, geographic-ethnographic approach, social-pedagogical approach, artistic-aesthetic approach, civilizational approach, social approach, philosophical-cultural approach, etc.

It is necessary to say that all of these approaches research particular branches of social-historical development.

Each of the approaches researches signs, processes, and changes relating to its research aims and mark them scientific and theoretical. In this case, it is good to remember the words of B. Russel, who studied human knowing activity. " Knowing," he wrote, "It mustn't be based on the ravine style between us and our ancestors which one cannot cross" [2]. "Knowing first, should be based on learning facts and second general relations among facts" [3]. Also, there exists "learning directed at mental activity" [4] and it relates to these learning behaviors. From these facts, learning can be understood as having subjective character, and learning directed at mental activity is considered objective or directed social-practical. In addition, the philosopher showed that "man's knowledge is not reliable, uncertain and limited" [5]. Ethnoculture is the social actuality expressing the existence of a nation, its lifestyle, language, customs, and traditions. By assimilating and changing the outer world and understanding himself, a people's or nation's material and spiritual wealth is created in the process of a long social-historical development.

Any social actuality, especially the culture relating all sides of human life, has several features and many complex, sineretic, universal, stability and changeable factors. Without taking into consideration these features of the culture, it is impossible to correctly understand its place in the social-historical development and the signs about inner instructions and functions. S. Shermuhammedov and

A. Ochildiyev wrote that, "society and culture aren't the same things but there exist culture in all spheres of the society. That's why without realizing the influence of society to the development of culture and on what shapes this influence acts, we can't define till the end the essence of a great social event, which is called culture" [6]. If we use this approach to ethnoculture, it has to connect culture retrospective with the forming process of the ethno, because there is not ethnoculture without ethno forming process. The origin of the Uzbek people, its ethnogenesis, has a very long past. As S. P. Tolstov wrote, "like all the people on earth, the Uzbek people had also been formed from different tribes and unites which were named with different names in the past and middle ages" [7].

M. Ermatov pointed out, "that the specific features of the Uzbek people's ethnoculture first gave an appearance in the primitive society system and appeared clearly in the next social-historical development stages. For example, half nomadic, nomadic and constant living manners are known as such kinds of signs" [8].

The interest in learning about the Uzbek people's ethnogenesis was at its highest point in the years of Independence. Summarizing the thought given in public information means and special publication, K. Sh. Shoniyozov clarifies:

> In Ethnography there are several signs marking household (nation too) such as: territory, language, economics, culture, common historical fate, thinking system, ethno's unity, being in the sphere of a certain state, having its ethnic name, common religion and others. But in fact, all of these signs couldn't be possible to play an important role or to exit at a certain time when a household was forming. If one or several of the ethnic signs appear at the same time the other may appear later.
>
> If a language plays the main role in forming of a certain ethno it is not far from the reality that a hosehold may be a leader mark for the second and a material culture for the third one [9].

So, ethnoculture is an important factor in ethnogenesis but all of its components are not seen fully at the same time. With taking an important place of the ethno in the social-historical development, the ethno-cultural signs would be seen more widely and express the peculiarities of the ethno more completely. One may learn the specific peculiarities in the Uzbek people's ethnoculture divided into the following categories: material and spiritual [10], social, artistic-aesthetic, morality, behaviour, gesture, daily life [11], and so on. S. A. Artunov, who investigated ethnoculture, divided the concept of ethnoculture into the spiritual culture that occurs in a human's microcosm and the material culture that occurs in a human's macrocosm [12].

Artunove also pointed out that there is another type of culture called physical or human, "the culture of body actions," which connects dialectically those two systems shown above [13]. These systems perform the normative communication functions which provide relations among the people and archaic connections like understanding the world and changing it, understanding the human being, and forming usage (from eating food to taking pleasure in literary aesthetic works) [14].

Certainly, classifying the ethnoculture into such systems and functions is based on certain scientific-theoretical conceptions and aims. In our opinion, it is not possible to limit the specific peculiarites of the ethnocultures with cultural signs appearing in primitive ages. Each social-historical development step affects the system and functions of the ethnocultures, and it is impossible to say a word about the culture of one people by looking at general and private (national, ethnic) signs without considering their social-historical context. But in the Uzbek people's ethnoculture, the aspiration for keeping their historical-cultural paradigms is strong. In our opinion, we should always think about strengthening these ethnic features in our people's culture and absorbing them into the soul of the young generation.

References

1. Yasters. K. The Meaning and purpose of history. –M.: Republic, 1998.- page 21.
2. Russel, Bertan. Man's thought-its spheres and limits. –M.: Foreign literature, 1957.-P.455.
3. The same page.

4. The same page.
5. The same page. –P.540.
6. Shermuhammedov.S. Ochildiyev.A. Culture and civilization. –Farg'ona, 2000. –P.26.
7. Tolstov. S. P. The Early culture of Uzbekistan. —T.: 1943. –P5
8. See: Ermatov. M. Ethnogeneses and forming of the Uzbek people. –T.: Uzbekistan. 1968. –P.8–9
9. Shoniyozov.Sh.K. Some theoretical matters about Uzbek people's ethnogeneses.// The Humanities in Uzbekistan, 1998, edition 6, p-33.
10. See: Philosophical encyclopaedia. –M.:1983. -P.292–295
11. See: Socolov.E. V. Culture and Personality. –L.: 1972.-P.67
12. See: Artunov.S. A. People and Culture. Development and Co-operation. –M.: Science.1989.-p.129.
13. See: The same page.
14. See: The same page.-P.130.

Understanding of Charity in Russia and its Interaction with the State

Maria Kislyakova

Saint Petersburg State University of Cinema and Television,
Saint Petersburg, Russia

Abstract. *This article discusses one of problems for Russian society – charity development. Its understanding is considered with regards to influence of charity on Russian society and its interaction with the state. The main problems and the search for their resolution are defined.*

Keywords: *charity, philanthropy, employment, volunteering, social policy.*

In Russia the charity is being developed in which large companies, foundations and people take part. According to the Law of the Russian Federation No. 135 *On Charity and Charitable Organizations* of August 08, 1995, a charity is defined as the activities of the voluntary citizens and legal entities related to the selfless (gratuitous or concessional) transfer of property, including money, as well as unselfish performance of works, provision of services, and other assistance to citizens or legal entities.

Charity is a type of activity that can be effectively used in the arrangement of social care and government institutions as well as non-state charitable organizations. The available tools are social projects, grants from grant-giving organizations, funder organizations, and private donations [1].

Charity is a wonderful Russian word, which is soft, melodious and nice. The literal meaning is completely clear – to do good and bonify.

For many people today, "social care" means a direct financial assistance or other type of assistance provided by wealthy people or organizations to the poor and others. Used as a synonym for the charity, philanthropy (from the Greek – "love to people") is an activity by which private resources are freely distributed by their owners in order to facilitate the people in need, to solve social problems, and to improve conditions for public life.

The concept of charity is multifaceted. Many people consider charity to be only cash donations, but in fact, this word has a deeper meaning. Many of us who do good deeds do not even think about our philanthropic activities. Indeed, charity, except for cash donations, is also expressed by the time people spend to help people in need as well as talent sharing by people with the community.

Charity is considered to be a moral action in response to the "human problematics". Charity for a long time has been a "social history of moral imagination." Charity is necessary for a free, open, democratic, and civil society. Charity completes different tasks; sometimes it tends to reduce the suffering and misery of people in need, to show mercy and compassion, and sometimes the main aim is to improve the quality of life conditions.

To date, charitable organizations have not gotten a very good reputation because of continuous ethical scandals. Mass media most widely reported about the scandal in the International Charity

Fund "United Way", the brightest representative of charitable activities for millions of Americans, connected with the abusive practice of non-profit organizations related to money laundering and involvement of such employees as disgraced lobbyist Jack Abramoff [2]. Reports also placed in doubts the ethical activities and decisions of the Red Cross and other charitable organizations participated in giving aid to the victims of the terrorist attack on 9/11.

At the same time, scientists predict a massive transfer of property in the coming decades – 41 trillion dollars in one opinion – that would start a "gold century of philanthropy". This growth and a high concentration of philanthropy, as well as scandals, led to the increased monitoring of charitable organizations, politicians, and regulators at the highest level. Of course, this debate deals with the control over charities, demanding a better understanding or justification of the primary activity as well as a better understanding of considerations such as exemption from taxes.

Kuban experience has shown that the adversity can unite people, who can be assumed to be the exact opposites: the activists of "Bolotnaya" and "Nashysty" show business representatives and voluntary electricians. Only a few volunteer groups worked in Krymsk (over 600 people) from the very beginning. There was a camp called "Volunteer" in the city center close to the administration building, "Dobryi Lager," and a camp with volunteers organized by the model Natalia Vodianova and many others who were located in the outskirts. An amazing outpouring of charity in response to the disaster – expressed in money, time, and organization – reminded millions of people around the world that they can perform their charitable role in the world as well. But these disasters made us think about the role of organizations such as the Red Cross and Doctors Without Borders versus the role of the government.

When the government was obviously distressed or even sharply criticized for the fact that it was not able to provide adequate assistance, does the Salvation Army have an equal responsibility for adequate needs satisfaction? Non-profit groups cannot be the only or even the main source of help in such cases, but what can or should be their role? What can they promote that government and business cannot? And should they be allowed to choose who they help and who they do not help? Such kind of issues we face even more, when we consider the growing role of NGOs (non-governmental organizations) staff in providing assistance and development efforts worldwide. After Putin visited Krymsk, many volunteer camps were dispersed by the local authorities. As a result, some of the volunteers left for home and others moved to another camp.

The "insufficient legal regulation of the volunteer activities" was discussed in Moscow before the events in Krymsk. This issue was considered as far back as in 2010 when volunteers extinguished massive forest fires. Developers of the projected law say that the main purpose of this document is to fight against tax evasion of nonprofit organizations that attract volunteers [3]. But it is not necessarily NGOs who have to attract volunteers; volunteers appeared in Krymsk themselves not from any organizations.

Charity is an ancient tradition, but the tradition in some danger. Each person should know something about the tradition of volunteering for the benefit of society.

Charity in the broad sense as we understand it penetrates our lives, whether we like it or not. Charity is as important in our lives as law and medicine, subjects about which we know much more than we know about charity. Philanthropy is an important tool in our collective attempts to solve social problems.

Irrespective of the place where we live, most of us took part in some form of voluntary action for the public good. But this does not mean that we are well-aware of such activity.

Despite the fact that charity is common in the culture, some people think about philanthropy very carefully. For example, many Americans believe that most of the financing comes from large foundations such as the Ford Foundation and from major corporations such as Microsoft. In fact, 83% of all charitable money in the USA is given by individuals, but not by corporations or foundations. Similarly, many people believe that most, if not all, of the funds obtained and distributed to non-profit organizations in the United States are from charitable contributions. In fact, only a small percentage of the revenues of the non-profit sector – only one of the eight dollars – comes from private philanthropy. In general, the

American non-profit groups receive less from the private provision than from the government, and their largest source of revenues is not the provision of private or government subsidies, but the fee for the services and products they render. [2]

Our opinion about charity is based on informal and often random sources; this is what we learned from the family, churches, and traditions. We did not study philanthropy like the economy or politics.

Unlike the other two major sectors of social life, namely, business and politics (or other major sector of private life, e.g. family), philanthropy has recently become a subject taken into consideration by scientists who have just recently begun to study it systematically.

Most of the activities that are marked as "charity" have existed for ages. Organized charity is older than democracy and capitalism and even older than Christianity and Buddhism, older than societies and many other traditions which do not exist anymore. Charity in its less organized and spontaneous form is as old as humanity itself.

Good deeds done by others, both in the past and present, make our life possible. People around the world are beneficiaries of scientific or medical discoveries funded by grants and charitable donations. One of the most troubling weaknesses of charity identification, such as voluntary return or help, is that too much attention is focused on the giver. This makes the wrong impression about the charity; it looks like the charity receives more than it gives.

All of us once needed help; by helping others we can better understand each other. We are all vulnerable. All of us were once babies. Most of us do not treat ourselves as "vulnerable" ones until one of us faces serious illness of our family and friends, or serious consequences of an accident that occurred while on vacation. In such situations we attach great importance to charity and giving assistance to strangers, and then forget about this.

Although philanthropy requires a relatively small portion of our resources – our time and money – its statistical data are still very impressive. Only volunteers are able to provide targeted assistance to those in need. Volunteers participated in the aftermath of natural disasters in Thailand, Haiti, and Japan; in 2005 over 13 thousand volunteers from all over the United States came by their cars and motorcycles to flodded New Orleans, while only several groups of over 600 persons worked in Krymsk from the very beginning.Charity is a force that features a great importance worldwide. Tens of millions of Americans give money in philanthropic way, sometimes because it is easier than giving their precious time and sharing their modest talent.

The total volume of charity in the world has decreased. At the same time, Russia in one year moved from the 130th to 127th place and was included in the top ten countries worldwide by number of volunteers. This is according to the World Giving Index study [4] which is held by the British Charity Fund CAF based on a worldwide survey performed by the Gallup Company. Over 155 thousand people from 146 countries took part in the 2012 rating. For the first time ever, the data on personal participation charity across the world over the past five years – from 2007 to 2011 – were analyzed.

According to RBC (Russian Business Channel), the country's place in the rating depends on the average value of three indicators: providing donations to charitable organizations, working as a volunteer, and helping the needy stranger.

According to the study, the growth of Russian charity was mainly caused by the increase in the number of people providing direct assistance to the needy persons – 36% in 2011 versus 29% in 2010. On average, 21% of Russians are involved in the charity, which is 3 percentage points higher than in 2010.

The United States was on the top of the ranking in 2011; they moved up from fifth place in 2010. They were followed by Ireland, Australia, New Zealand, and the United Kingdom. Also in the top ten were: the Netherlands, Canada, Sri Lanka, Thailand and the Lao People's Democratic Republic.

Russia has moved from its 130th to the 127th place due to the following indicators: 7% of charitable donations (5% – in 2010, 6% – in 2009), 17% of volunteer work (23%, 20%), 29% of assistance provided to the needy people (36%, 29%). On average, 18% of the Russian population is involved in charity, which is 3% less than it was last year – an increase in donations was covered by the decrease of other indicators. Taking the higher place by Russia in the ranking was mainly caused by the fact that charity in many countries has deceased even more dramatically.

The number of volunteers in Russia hit the ten top with its 8th place ranking. 21 million people were engaged in volunteering in our country in 2011.

However, it is not all that optimistic: the growth of charitable activity in the world is due to the fact that people started to provide more assistance to persons in need and to perform volunteer work, while the amount of donations to charitable organizations has reduced. The number of those who provide assistance to the people in need has increased by 2% and the number of volunteers by 1%. The number of respondents who declared that during the last month they had donated money to organizations fell by 1%.

"One of the tragic consequences of the financial crisis in many countries deals with a decrease in the amounts and reduction of the number of donations," says John Low, CAF Executive Director, who commented the rating results. "It means that there will be less aid rendered to victims of natural disasters and catastrophes, fewer people will have access to water, personal hygiene and proper living conditions, there will be less opportunity to care about sick people, the elderly and children". [5].

Those whose life is going well were always in charge of the care for the public good by helping people in difficult situations. For example, charity in the modern United Kingdom is an integral part of everyday life. There are over 9 thousand different organizations in existence due to the citizens' donations in the country. About 70% of the UK population regularly gives money for charity purposes. British philanthropists take care not only of the needs of their fellow nationals, but also of the citizens of other countries who have to cope with the consequences of natural disasters, famine, lack of medicine, drinking water, etc.

A charity's scope is only one measurement tool to assess its significance. Its diversity and the extraordinary is another thing. As noted above, charities are considering the most important social and moral problems influencing society as well as our individual lives. In fact, the most important issues facing society – moral issues such as social welfare, human rights, and the environment – often occur first of all in the "third sector", which is a public space where volunteer work is carried out by the society.

Charity has a significant impact on social, political, religious, moral, economic, scientific, and technical matters. The range of issues being protected by charitable organizations includes ones from efforts to limit air pollution up to determinations of the rights of children, as well as from providing opportunities for artists' exhibitions and funding hospitals for the care of the terminally ill people. Charity had an impact on shaping solutions with respect to religion, education, health, social welfare and social services (including family, children and youth), arts and human sciences, culture preservation, social works, sports and leisure, international aid and development, and the environment.

Philanthropic methods are as varied as the needs which they satisfy. Food and drinks, companionship and compassion, medicine, exemption, work, education, religion, and music are all the needs which are completely satisfied by the charity giving volunteer gifts, such as money or services. The actions available are dictated by the needs.

However, in assessing the scope of charity, it is worth noting once again that there is a huge and largely unknown "ocean" of the informal, spontaneous, and interpersonal philanthropy. We can make a mistake assessing the scope of the charity sphere if we neglect or forget about the all-permeating good deeds, which are immediate, direct, or personal, the traditional domain of welfare addiction, love of neighbor, courtesy and tolerance, and "ordinary virtues" if there are any.

The essential feature of today's life is expanding the circle of persons representing the vulnerable segments of society. They include aged and old people, unemployed persons, migrants, disabled ones, chronically ill people, persons who are below the poverty line, beggars etc. Increasing the number of such persons adversely affects the whole society, causing economic, physical, and moral damage. Therefore, it is important that society realizes the practical value of charity as a tool for the social rehabilitation for truly needy citizens and reducing the severity of health problems. A diseased tissue of human mutual relations is being restored through mercy and compassion.

Today, it would be useful to refer to the experience of our predecessors, whose charitable activities were dictated by the realization of human solidarity and the desire to eliminate the risk associated with people in need in the society.

The successful operation of charitable organizations relies on some basic facts – first of all improving the legal framework for activities of these organizations, networking between private charitable organizations, and governmental structure of social protection.

There is a growing demand for well-trained professionals in this sphere. In the future it is also necessary to create clear, workable operation programs for charitable organizations. Another important factor is to make the social service customer be aware of the fact that he can be helped by the charitable organization.

References:

1. D. Bell, Social framework of the information society. New technocratic wave in the West. - M: Progress, 1986. – page 121. (In Russian)
2. Understanding Philanthropy Its Meaning and Mission, Robert L. Payton, Michael P. Moody, Bloomington, Indiana University Press, 2008, page 224
3. "Money", Economic Weekly, the publishing house "Kommersant", No.29 [886] 23.07-29.07.2012 (In Russian)
4. Charities Aid Foundation 2011
5. „New Region", NR2.Ru, 2011

Political Consciousness of the Youth as a Foundation for the Development of Civil Society

Eleanora Yusupova
*Tashkent State Pedagogical University named after Nizami,
Tashkent, Uzbekistan*

Abstract. *This article is about the importance of formation of political consciousness of young people in terms of independence of the Republic of Uzbekistan.*

Keywords: *Political consciousness, national idea, educational policy.*

Modern educational systems represent an integral component of the culture of the society. The value of education is largely determined by the fact that it is not only a basic prerequisite for the development of human civilization, but also a spiritual foundation and support for personality. High dynamics of modern social development predetermines the role and importance of the educational system in all spheres of human activity. Modern society requires from a citizen maximum demonstration of intelligence, creative abilities, and sensitivity to the current political, economic and social changes, high social mobility and psychological stability.

Various qualities of a comprehensively developed person necessary for full-fledged human activity are developed in the process of socialization. Period of the most intense accumulation and development of common cultural, political, and professional knowledge, the development of a person as the individual, his (her) formation as a citizen takes one of the most important places in this complex and contradictory process. For a large social group — the youth this is the period of learning in a higher vocational educational institution. In modern conditions of formation and development of society the importance of education in general and higher education in particular grows up more and more. And here, along with the need to prepare a highly qualified professional specialist, the task of forming a politically educated person corresponding to the changed characteristics of the domestic political system is brought to the forefront. In this regard, the need for objective scientific study of the effect of the higher vocational educational system as a specific institution on the functional aspect of political culture that is political consciousness of young people appears [3, p.27].

It is known that the main task of the educational policy of any state is the creation of such educational system which, on the one hand, would reflect socially, mentally and culturally the economic and political needs of a particular stage of society and state development, and on the other hand would be a powerful tool, the institute allowing to provide the relative political and social unity of society, its socio-political stability, especially in the conditions of deep differentiation.

Educational system acquires the character of one of the state and society major political institutions which aims, inter alia, at the formation of "political man", i.e., the formation of his (her) political qualities as a citizen. In this regard, among other forms of social consciousness experiencing the impact of the educational environment the political consciousness occupies a special place. It is the political consciousness that the most vividly reflects the attitude of a person to the socio-political, economic and other processes, taking place in society, and contributes to the formation of the political behavior of this person.

The youth active in politics studies, analyzes youth problems, and is able to solve them, entering the representative and executive authorities in a small amount yet. Young people must be convinced that the effectiveness criterion of any policy is the growth in prosperity of each and every one providing, eventually, wealth, fame and power of the state. The youth is able to become one of the main driving forces in building a civilized civil society.

The Republic of Uzbekistan began handling the problem of increasing the political activity of citizens with a legal confirmation of their rights to participate in state administration. The Constitution proclaimed legal guarantees of inclusion of a human being in the system of his (her) relationship with civil society and the state. Citizens of the Republic of Uzbekistan have the right to participate in administration of public affairs and the state affairs both directly and through their representatives. Such participation is carried out via self-administration, referendums and democratic formation of government bodies. [1, p. 8]. Formation of principles of civil society and institutions of Uzbek democracy became one of the stages of reforming the political system, the most important feature of which is the growth of political consciousness of the people.

In the present period, when the society and economy reforming is being updated, as well as the state is being formed, we cannot lose sight of our younger generation, since it is the one that shall have to walk further along the path to democracy. For this purpose, we need to pay more attention to the education of the youth in the spirit of the time, taking into account all innovations. Political consciousness is directly related to the ideology of the state, so now a great deal of attention is paid to the ideological sustainability of our youth. It is very important to form in our people, especially the younger generation, the ideological immunity and political knowledge base in the current context of overcoming the ideological vacuum. The emerging idea of national independence, based on age-old traditions of the people and human values, should clearly reflect the high goals and objectives of society and the state [2]. Further deepening of the processes of society and the state democratization, ensuring their consistency and effectiveness involves the following:

Firstly, the liberalization of the political life of the country, the state and social development, increasing of political activity of citizens, the formation of a political culture based on national and universal values, ensuring standards of democracy, freedom of thought and conscience, pluralism and human rights, priority of the principle of "living in accordance with the humanistic and universal values" [4].

Secondly, the establishment of an effective mechanism, ensuring a balance between the interests of the social forces and movements existing in society, as well as approval of multiparty politics principle in the political life.

Thirdly, the creation of conditions for independent democratic institutions, strict observance of constitutional principles of separation of powers; the creation of maximum opportunities for implementation of political and social potential of freedom and initiative of members of society.

Fourthly, the expansion of the activities of local authorities and autonomous bodies of citizens, step-by-step transfer of a part of the powers of governmental authorities. This means that as far as political consciousness and participation of people is growing, the political culture is developing a number of the state functions shall be transferred to the autonomous bodies.

Fifthly, the improving of the system of selection of talented, well-rounded and highly qualified, and devoted to the Motherland young professionals capable to carry out reform tasks in accordance with democratic requirements, as well as the interests of society.

All this is a very difficult and complex process, which has always flowed with difficulties. It requires a change in thinking and outlook of people, elimination of subjectivity phenomena, regionalism and other vices of bureaucratic system. Therefore we need a harmonious and comprehensively developed youth, politically well grounded, able to take control of the state in the future into their own hands. Level increase of political consciousness is one of the important aspects of the educational process within the youth. None of the developed state could exist without the political culture and political consciousness of its citizens. Political culture empowers the youth with such a wealth of knowledge, skills, and cultural values, without which there is no future for this state; as a matter of fact the youth is this future.

References

1. Konstitutsiya Respubliki Uzbekistan. Tashkent, 2012. [The Constitution of the Republic of Uzbekistan] (in Russian)
2. Karimov I. A. Ideologiya — eto obyedinyayushchiy flag natsii, obshchestva, gosudarstva. Tashkent, 1998. [Ideology is a unifying flag of nation, society, state] (in Russian)
3. Korenev I. Osnovy formirovaniya novogo politicheskogo soznaniya obshchestva // Pravo, 1998. No. 4. [Basis of the formation of a new political consciousness of society] (in Russian)
4. Nishanov M. N., Dzhavakova K. V. Politologiya. Tashkent, 2005. [Political Science] (in Russian)